Mathematical Puzzle Tales

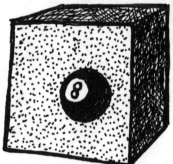

Martin Gardner

Mathematical Puzzle Tales

Martin Gardner
With a Foreword by
Isaac Asimov

For Isaac Asimov,
of course

Inquiries should be addressed to: The Mathematical Association of America, 1529 Eighteenth Street, NW, Washington, DC 20036.

This book was published previously as *Science Fiction Puzzle Tales* by Clarkson N. Potter, Inc., in 1981.

Library of Congress Catalog Card Number 00-109558

ISBN 0-88385-533-X
Printed in the United States of America

Current Printing (last digit):
10 9 8 7 6 5 4 3 2 1

CONTENTS

ANSWERS

FOREWORD

In my lengthening life, I have been fortunate enough to meet a number of rational men—not a large number, to be sure, but enough. One rational man makes up for thousands of fuzzy thinkers when it comes to companionship.

Of all those I have met, Martin Gardner is the quietest and least impassioned about it, but very effective just the same. Back in the 1950s, he wrote his classic *Fads and Fallacies in the Name of Science*, and there has never been a harder blow at science's irrational fringe. The quietness of Martin's style made it difficult for anyone to inveigh against the book, and its rigid rationality deprived them, in any case, of any sensible grounds for doing so.

Most people who encounter Martin these days do so in the pages of *Scientific American*, where, for a quarter of a century, he has composed a monthly column called "Mathematical Games," and done so with remarkable insouciance and charm.

So delightful is his approach, that on many occasions I have found myself reading the column with pleasure, though some of its details have managed to get past me. Even when you can't quite see the details of the corner of mathematics that Martin spreads out before you, the grand design is unfailingly fascinating.

It might be supposed that one should sneer a bit at anything called "recreational mathematics" or "mathematical games." Are such things not just "recreations"? Just "games"? They can't be of any importance, one might think. They're just a way of fooling around.

Who really cares how many ways you can pair up people at a bridge table, or how many colors are needed to fill up a map under certain conditions, or how many routes a knight can take about a chessboard, or what the shortest path from one city to another might be if you follow a particular type of route?

But mathematicians do care and always have.

It might be argued, in fact, that all mathematics begins

as "Games" in the sense that the first glimmerings of any part of it seem to have no use.

There must have been a time when some prehistoric genius said to a friend, "See here. Suppose I have two stone knives. I can make two piles out of them, both with the same number. If I add another knife, I can't divide them equally. If I add still another knife, I can. And if I add yet another knife, I can't. Do you suppose this sort of thing goes on forever?"

Undoubtedly, the friend said, in honest astonishment, "Who cares? Why are you sitting there dividing knives into heaps? Use one of them to kill something. Do something practical!"

Also, undoubtedly, the primitive mathematician found it wonderful to continue to pass the time considering this question of equal piles and to wonder further if there were any system to the matter of making three equal piles and so on.

It was just a game; it had no practical use. Eventually, though, such questions of divisibility—of the behavior of heaps when combined or divided into unequal heaps, or into arrays—came to be generalized into systems of computation that made it possible to add, subtract, multiply, and divide.

Imagine the excitement when a government official discovered for the first time that such computations would make it far easier to collect taxes and keep track of expenditures. At once the game stopped being a game and became a hard-nosed business, very suitable for "practical" men.

There will always be a tendency, however, for mathematicians to move away from those branches of the field that are too readily applied to the daily business of life. Unquestionably, there is more fun in fiddling around with problems when no one is glancing over one's shoulder and saying, "Have you got it worked out yet? We desperately need it to turn a profit this quarter."

Yet it is so hard to keep the games recreational. Many a time mathematicians have been convinced they had some kind of problem that couldn't be of use to anybody, something so massively unimportant that no one could conceivably want to interfere with the mathematical pleasure of such nonsense. Then someone comes along and finds that this "nonsense" can be used to make possible intricate telephone-switching capabilities or to explain the behavior of arcane subatomic particles. Mathematicians must then find another refuge.

Well, Martin offers everybody (not just mathematicians) creative refuge for the imagination. The puzzles in this book are not just puzzles. Very often, they embody deep mathematical principles that deal with matters not yet well enough understood to be applied to the practical world. Such "games" are *not* more trivial than "real" mathematics. They may well be more important and may be the foreshadowing of future mathematics.

The puzzles that follow are woven into short science fiction stories. They add amusement but are not the essence of the book. The science fiction is important, though, because it demonstrates that however times, customs, and technologies alter, the essence of mathematical relationships is, was, and will be the same. It is probably the only truly rigid and relentlessly constant factor in an otherwise ever-changing universe.

Isaac Asimov
January 1981

PREFACE

When Isaac Asimov and George Scithers began planning
Isaac Asimov's Science Fiction Magazine (henceforth,
IASFM) in 1976, George approached me at a gathering of the
Trap Door Spiders, a curious club to which the three of us
belong, to suggest a puzzle feature. Was it possible, he
asked, to weave a puzzle into some sort of science fiction
vignette or pastiche? In other words, present the puzzle
with a SF story line. If so, would I be interested in doing this
as a regular feature for the new magazine?

The idea was intriguing, especially since I had once
perpetrated two SF stories based on topological curiosities:
"The No-Sided Professor" and "The Island of Five Colors."
I cleaned up an off-color combinatorial problem, never before published, for my contribution to Volume 1, Number 1,
dated Spring 1977, and have been writing the puzzles ever
since. I enjoy writing them and I know from letters that
readers like to work on them.

This book brings together the first thirty-six *IASFM*
puzzles. To almost all of them I have added a postscript,
which allows me to explain some (not all) of my compulsive
wordplay, to thank whoever should be thanked, to discuss
feedback from readers, and to suggest books and articles
containing interesting material relating to the puzzles.

Good puzzles are usually jumping-off points for serious
mathematics. You'd be surprised how much math you can
learn by exploring some of the implications and ramifica-
tions of what may seem at first no more than a trivial brainteaser.

HOW TO USE THIS BOOK

There is always a strong temptation, when answers are given in a book of puzzles, not to spend much time trying to solve a puzzle. It's too easy to turn to the answer!

To get the most fun and instruction out of this book, let me urge you to work on every problem before you give up and check its answer. Each story-puzzle is numbered; the solution (or solutions), similarly numbered, will be found in the First Answers section. In most cases a new problem, related to the original, is posed at the end of the answer. These are solved in the Second Answers section. In some cases there is then a third question at the end of the second answer that leads to still another solution in the Third Answers section.

At the end of almost every final answer you will find a postscript in which there are additional comments concerning the preceding questions. Occasionally I mention books and articles of special interest that will tell you more about the topics under discussion.

To make for easier reading, only the titles of books are given. These books, including my own, are listed alphabetically by author, with publisher and date of the most readily accessible edition, in the bibliography at the back of the book. Periodicals and journals receive full citations in the text.

INTRODUCTION TO THE MAA EDITION

Three book collections have been made of the whimsical columns that I wrote for *Asimov's Science Fiction Magazine*. This was the first. The second anthology, *Puzzles from Other Worlds* was published by Random House in 1984. The third, *Riddles of the Sphinx*, was published by the MAA in 1987 in its New Mathematical Library series.

Ronald Couth, the computer scientist on board *The Bagel*, spoofs the name of Donald Knuth, Stanford University's famous computer expert. *The Bagel* is, of course, a take-off on Darwin's voyage on *The Beagle*. Astute readers will spot other word plays on proper names.

For this edition of *Science Fiction Puzzle Tales* I have added a Postscript to update some of the chapters.

Martin Gardner

PUZZLES

1

LOST ON CAPRA

Dr. Ziege, the eminent German extraterrestrial geologist, was the first human to set foot on Capra, the fifth planet from the star Capella. For several months she and her two companions explored the planet by spacecar.

Capra is roughly twice the size of the earth, but lacking in enough water to support life. Dr. Ziege found the planet a barren, sandy waste, its surface as smooth as the plains of Kansas. Like the earth, Capra rotates on an axis. Dr. Ziege designated one pole north and the other south in conformity with the ship's magnetic compass and the planet's earthlike magnetic field. Geographic and magnetic Caprian poles coincide.

The last radio message from Dr. Ziege was: "We have lost our bearings and cannot find the spaceship. Yesterday we drove 10 myriameters due south from our last camp site, then 10 myriameters due east, then 10 myriameters due north. We find ourselves back at the camp site. Food supplies exhausted. Send help."

Attempts to reach Dr. Ziege for precise information as to her location brought no response. The German government immediately fired a rescue ship through Wheeler wormhole 124C41+. Two days later it was circling Capra with plans to land near the north pole. It seemed obvious that only from that pole could Dr. Ziege and her men go 10 myriameters south, then east, then north and be back at the starting spot. But there were no signs of the explorers within a radius of 20 myriameters from the north pole.

"Ach!" shouted Felix, striking his temples. "We are looking in the wrong place. *Another* spot fits Dr. Ziege's message perfectly."

"How can that be?" said Hilda. "If the starting spot is a few kilometers from the north pole, the terminal spot will miss the pole by a short distance. The farther south you go, the more it misses. At the equator it misses the starting spot by a full ten myriameters. And south of the equator it will miss by more than that!"

3

Nevertheless Felix was right. Where should they look next?

2

THE DOCTORS' DILEMMA

The first earth colony on Mars has been swept by an epidemic of Barsoomian flu. The cause: a native Martian virus not yet isolated.

There is no way to identify a newly infected person until the symptoms appear weeks later. The flu is highly contagious, but only by direct contact. The virus transfers readily from flesh to flesh, or from flesh to any object which in turn can contaminate any flesh it touches. Residents are going to extreme lengths to avoid touching one another, or touching objects that may be contaminated.

Ms. Hooker, director of the colony, has been seriously injured in a rocket accident. Three immediate operations are required. The first will be performed by Dr. Xenophon, the second by Dr. Ypsilanti, the third by Dr. Zeno. Any of the surgeons may be infected with Barsoomian flu. Ms. Hooker, too, may have caught the disease.

Just before the first operation it is discovered that the colony's hospital has only two pairs of sterile surgeon's gloves. No others are obtainable and there is no time for resterilizing. Each surgeon must operate with both hands.

"I don't see how we can avoid the risk of one of us becoming infected," says Dr. Xenophon to Dr. Zeno. "When I operate, my hands may contaminate the insides of my gloves. Ms. Hooker's body may contaminate the outsides. The same thing will happen to the gloves worn by Dr. Ypsilanti. When it's your turn, you'll have to wear gloves that could be contaminated on both sides."

"Au contraire," says Dr. Zeno, who had taken a course in topology when he was a young medical student in Paris. "There's a simple procedure that will eliminate all risk of any of us catching the flu from one another or from Ms. Hooker."

What does Dr. Zeno have in mind?

3

SPACE POOL

PUZZLE

Two young physicists were discussing their vacation plans.

"I may take a space cruise," said Jones. "I've been told that the food and the girls on the *Cutty Snark* are superb, and that this summer the cruise includes landings on the moon, Mars, and Venus."

"I went last year," said Smith, "and had a marvelous time. The ship has a huge recreation room with all sorts of new games. Space pool, for instance. When the ship's in a g-field it's played the regular way, only the table is enormous and there are more than 100 balls."

"How is it played in zero gravity?"

"Some engineer figured out a way to create a green-tinted magnetic force field," said Smith. "It keeps the balls inside a rectangular parallelepiped about a meter above the table. The ivory balls have iron cores. They bounce off the green walls the same way they bounce off the cushions on the table. The wooden cues are not affected by the field, so you can poke them into the field at any spot. The pockets are holes in the field's eight corners. If a ball hits a corner it leaves the field and you score the ball's number like in ordinary pool."

"But won't the balls keep on moving after they're hit? How can you stroke the cue ball when it's on the wing?"

"The balls freeze exactly ten seconds after each stroke," said Smith. "I don't know how it works. I think another force field brings all the balls to a dead stop."

"How many balls are there?"

"I can't recall. Somewhere between one and two hundred. When the game's played on the table it starts with the balls packed into a triangle like the 15 balls of regular pool. When it's played in space you start with the same set of balls packed into a regular tetrahedron."

"In other words," said Jones, "the number of balls is both triangular and tetrahedral. There can't be many numbers like that."

Smith closed his eyes. "Well, there's 1. It's triangular

6

and tetrahedral, but that's trivial. The next tetrahedron is a triangle of 3 balls with 1 ball on top, or 4 altogether. But 4 balls won't make a triangle."

"Ten will," said Jones. "It makes a triangle with rows of 1, 2, 3, and 4. And it also makes a tetrahedron. Every tetrahedral number is the sum of consecutive triangles; and triangles 1, 3, and 6 add to 10."

Jones took out his calculator. "Let's see. If I remember my number theory, triangular numbers have the form $\frac{1}{2} n(n + 1)$ where n is any positive integer. Tetrahedral numbers have the form $\frac{1}{6} n(n + 1)(n + 2)$."

It didn't take Jones long to discover that the third number to fit both formulas was between 100 and 200. He could find no other solution less than 200, so this was the number he wanted.

With the aid of a pocket calculator, how quickly can you determine the number of balls (not counting the cue ball) used in space pool?

4

PUZZLE

MACHISMO ON BYRONIA

Byronia, a small planet that orbits a sun near ours, has a humanoid population similar to our own. The most striking difference is that Byronians come in three sexes. They correspond roughly to what we call male, female, and bisexual.

Because bisexuals have both male and female organs, they can perform as either sex and also bear children. Whenever a "mother" (female or bisexual) gives birth, the probability that the child is male, female, or bisexual is exactly one-third for each.

The new Supreme Ruler of Byronia, Norman Machismo, is a virile, hot-tempered male who gained total power by defeating a rebellious army of bisexuals. To solve the "bisexual problem" Machismo has issued the following decree: Every mother on Byronia, as soon as she or it gives birth to a bisexual, is to be rendered incapable of further conception.

Machismo reasoned like this. Some mothers are sure to have two, three, four, or even more heterosexuals before having a bisexual. True, occasionally a mother will have a bisexual first child, but that will be the end of her childbearing, so these births will contribute only a small percentage of bisexuals to the population. In this way the proportion of bisexuals in the population will steadily diminish.

Will the Supreme Ruler's plan work?

Shurl and Watts, at a base on Pluto, are in charge of distributing doyles to more distant outposts. Doyles are the size of peas, all identical, each weighing precisely 1 gram. They are indispensable in hyperspace propulsion systems.

Doyles come in cans of 100 doyles each, and shipments are made up of six cans at a time. The Pluto base has a sensitive spring scale capable of registering fractions of milligrams.

One day, a week after a shipment of doyles, a radio message came from the manufacturing company in Hong Kong. "Urgent. One can is filled with defective doyles, each with an excess weight of 1 milligram. Identify can and destroy its doyles at once."

"I suppose," said Watts, "we'll have to make six weighings, one doyle from each can."

"Not so, my dear Watts," said Shurl. "We can identify the can of defectives with just *one* weighing. First we number the cans from one through six. Then we take 1 doyle from the first can, 2 from the second can, 3 from the third, and so on to 6 from the sixth can. We place this set of 21 doyles on the scale. It will weigh n milligrams over 21 grams, and of course n will be the number of the defective can."

"How absurdly simple!" exclaimed Watts, while Shurl shrugged.

A month later, after the next shipment, another message arrived: "Any of the six cans, perhaps all of them, may be full of defective doyles, each 1 milligram overweight. Identify and destroy all defective doyles."

"This time," said Watts, "I suppose we'll have to weigh separately a doyle from each can."

Shurl put his fingertips together and gazed at a picture of Isaac Asimov on the wall. "A capital problem, Watts. No, I think we can still do it in just one weighing."

What algorithm does Shurl have in mind?

6

THE THIRD DR. MOREAU

It is not widely known among science fiction buffs that Dr. Moreau, about whom H. G. Wells wrote his famous SF novella, had a grandson who calls himself Dr. Moreau III. Dr. Moreau III is a professor of genetics at King's College, in London, where he is considered one of the world's top researchers in genetic engineering.

By tinkering with a microbe's DNA helix, Dr. Moreau III recently managed to produce a strange new type of one-celled organism which he calls *Septolis quarkolis*. Drawing nourishment from the air and using energy derived from quarks, the new microbe splits every hour into seven replicas of itself. Each replica instantly becomes the same size as the original. Thus after 1 hour a single microbe becomes 7, after another hour the 7 become 49, in another hour the 49 become 343, and so on. As Dr. Moreau put it, in his report in *Nature*, *Septolis quarkolis* "multiplies at an alarming rate."

One day Dr. Moreau III put a single microbe, just "born," into a large and empty glass container. Fifty hours later the container was completely filled. Dr. Moreau then quickly destroyed all the microbes by bombarding them with tachyons. Otherwise, in a few more days they would have engulfed all of King's College.

I happened to be visiting King's College a few days after this event. Always alert for puzzle possibilities, I asked Dr. Moreau III's assistant, a chimpanzee named Montgomery, when the glass container was exactly 1/7 full of the microbes. Montgomery whipped out his pocket calculator and started working on the problem, but an hour later he still didn't have the answer. It was, he told me, beyond the capacity of his computer's readout.

Can the reader determine how many hours elapsed until the container was 1/7 full?

7

THE VOYAGE OF THE *BAGEL*

The *Bagel*, a huge spaceship shaped like a torus and rotating to provide artificial gravity, has just begun acceleration toward the center of the Milky Way. Its mission is to determine if the galactic center is a black or a white hole. The crew consists of five hundred men and women. The time is the middle of the twenty-first century.

Now that the initial jubilation over the start of the voyage has settled down, two mathematical physicists, Leo and Ling, are having supper. Leo is doodling on his napkin. Suddenly he bangs his fist on the table.

"By Asimov, I've got it! While everyone was getting introduced to each other in the last few weeks, the number of persons who have shaken hands an odd number of times is even."

"That's ridiculous," says Ling.

"No," says Leo. "It's a perfectly general theorem. In any group of people the number of them who have shaken hands an odd number of times, with members of that same group, is even."

Can you prove Leo's theorem?

8

PUZZLE

THE GREAT RING OF NEPTUNE

Captain Quank, *Bagel*'s top navigation officer, was busy penciling diagrams. Occasionally he punched the keys of his desk calculator.

Quank's task was to obtain information on the planet Neptune. To the crew's astonishment, they had found the planet surrounded by one enormous ring of low-density dust. It was no more than a centimeter thick—totally invisible in earth's telescopes.

The ring was bordered by two perfectly concentric circles. The ship had cruised above the ring on a straight line that crossed the outer circle at A, was tangent to the inner circle at B, and crossed the outer circle again at C.

"We know that distance AC is 200,000 kilometers," said Captain Quank. "The question is—what's the ring's area?"

"Won't we need to know the radii of the outer and inner circles?" asked Lieutenant Flarp.

"We'll get that information eventually," said the captain. "But now we don't need it. According to a curious theorem—I remember it from an undergraduate course in Euclidean geometry—the area of the ring is uniquely determined by this chord AC."

"You mean," said the lieutenant, "that given the length of AC, the ring's area is a constant regardless of the sizes of the circles?"

"Right! It's hard to believe, but it's true. I'm trying to remember how to calculate the area."

First question: What's the area of Neptune's great ring?

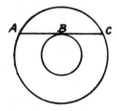

It was not until the third decade of the twenty-first century that Dr. David Klonefake at GIGE (Geneva Institute of Genetic Engineering) succeeded in producing a one-celled marine organism, almost microscopic, with a form like a torus. A torus is the topologist's term for a surface topologically the same as that of a doughnut. The animals are called *Toroidus klonefakus*, more commonly known as toroids.

THE TOROIDS OF DR. KLONE-FAKE

Toroids reproduce from buds that grow on their surface. The buds quickly enlarge to flexible rings which then break away from the "mother" and swim about by means of hundreds of tiny flagella that cover the surface like fine hair.

A fully grown toroid is not normally linked to its mother, but sometimes it grows in such a way that it is permanently interlocked with the parent or with a sibling. It is not uncommon to find three toroids linked in the curious manner shown in Figure 1, known to the topologists as Borromean rings. Study the picture and you will see that no two animals are interlocked, yet all three are linked.

FIGURE 1

If one is eaten by a larger life form, the other two become free of each other.

Dr. Klonefake's assistant had been observing toroids for months under a huge magnifying lens, trying to classify all the distinct ways they become linked. "This is incredible!" she exclaimed one morning. "Here's a colony of about fifty toroids, linked in a circular chain like a necklace. There's no

13

way one of them can break loose. But if any toroid is eaten, all the others instantly are unlinked!"

Dr. Klonefake couldn't believe it until he came over and saw for himself. Can you conceive of how the toroids were linked to one another?

When Philo Tate became postmaster general of the U.S. colony on Mars, he saw a chance to fulfill a lifelong ambition. Why not, he said to himself, issue a series of stamps with carefully chosen values so that no more than three stamps would be needed to give a total value for any positive integer from 1 through whatever number would be the highest possible?

At that time on Mars a postcard cost one dollar to mail from any dome in the colony to any other dome. The lowest stamp in such a series obviously had to have a value of 1. Suppose the series consisted of just two stamps. To meet Tate's conditions, the best choice of values are 1 and 3. It is easy to see that by using one, two, or three stamps one can obtain any sum from 1 through 7:

$$1 = 1 \qquad 3 + 1 + 1 = 5$$
$$1 + 1 = 2 \qquad 3 + 3 = 6$$
$$3 = 3 \qquad 3 + 3 + 1 = 7$$
$$3 + 1 = 4$$

No other choice of values for the two stamps can give all consecutive values from 1 through 7 or any higher number.

It is not hard to solve the problem for a series of three stamps. The best one can do are the values of 1, 4, and 5. By using one, two, or three of these stamps one can obtain any sum from 1 through 15 dollars.

Philo Tate's first issue was a set of four stamps. The dollar stamp was a maroon portrait of Edgar Rice Burroughs. The other three featured the faces of H. G. Wells, Ray Bradbury, and Isaac Asimov. By choosing the values carefully, Tate was able to obtain all sums from 1 through 24, and without using more than three stamps.

What values were assigned to the Wells, Bradbury, and Asimov stamps?

By the way, when the first supply of the stamps arrived from Washington, D.C., Tate said to his assistant, a young lady who had been born on Mars and lived there all her life, "Let's get these distributed to our branch offices as soon as possible." Then he remembered an old verbal joke he used to play on postal employees when he first started to work, as a young man, in the New York City post office. "As we used to say back in old Manhattan," said Tate, "neither rain nor snew can detain our couriers from their appointed rounds."

"What's rain?" she said.

11

CAPTAIN TITTLE-BAUM'S TEST

Captain Oscar Tittlebaum, in charge of the earth's first space station—it was located midway between the earth and the moon—was constantly annoyed by how often his crew members failed to understand what he believed to be clear instructions on the orders and memos he issued from time to time. To test each man and woman's ability to follow simple directions, he prepared the following test. You are invited to take it to see how you would have scored had you been a member of the space station crew.

THE TEST

Before starting this test, read all the instructions carefully. Use a pencil or a pen to enter all required information in the spaces indicated. *No erasure of any answer is permitted.* However, the test has no time limit, so proceed slowly, and follow all instructions to the letter.

Instructions:
1. In the six squares below print the letters in the last name of the man who formulated the three laws of robotics.

☐ ☐ ☐ ☐ ☐ ☐

2. In the nine squares below print the initial letters of the known planets of the solar system, starting with the initial of the outermost planet in the first square, then proceeding in sequence to the planet nearest the sun.

☐ ☐ ☐ ☐ ☐ ☐ ☐ ☐ ☐

3. Cross out just seven letters in the statement below so that what remains will still express a sum of 18.

FIVE PLUS SIX PLUS SEVEN

17

4. Draw a circle around the verb that does not belong in the following set:

BRING BUY CATCH DRAW
FIGHT SEEK TEACH THINK

5. In the square below put the number of the century in which July 4, of the year 2000, will occur.

6. In the space below print the shortest name (fewest letters) of a U.S. state that shares a letter in common with each of the other forty-nine states.

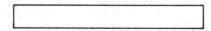

7. What is the length of line x in the triangle drawn below? Print this length in the square on the right.

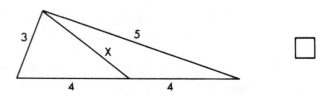

8. In the six squares below print a three-letter abbreviation for one month, followed by a three-letter abbreviation for another month, to produce a six-letter English word.

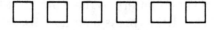

9. Place a prime digit in each of the three squares below. No two digits must be alike, and 0 and 1 are, of course, ex-

cluded. The three-digit number that results must be a multiple of each of the prime digits.

☐ ☐ ☐

10. At noon and at midnight the long and short hands of a clock are together. Between noon and midnight, how many times does the long hand pass the short hand? Print the answer in the square below.

☐

11. Write on the line below: "I always follow instructions carefully."

_____

12. Ignore all previous instructions with the exception of the test's first paragraph that begins with "Before" and ends with "letter." Sign your full name below and give your copy of the test to the captain.

_____

12

EXPLORING CARTER'S CRATER

PUZZLE

On Mercury there is a crater which, for this problem's purposes, we assume to have a perfectly circular rim. It is called Carter's Crater after President Jimmy Carter's great-grandson, who became an astronaut and was the first to set boot on Mercury. Carter landed near the crater that now bears his name. At two random spots on its rim he established supply stations. Although Carter returned to earth safely, an unfortunate fire on the spaceship destroyed his records of where the supply stations had been placed.

Two years later, astronauts Smith and Jones were sent on a mission to explore Carter's Crater. They landed at a random spot near the rim, picked a direction (clockwise or counterclockwise) by flipping a flat piece of Mercury rock, and started walking along the rim to the nearest supply station.

"Assuming that the two stations, and the spot on the rim where we started walking, are all randomly and independently selected points," said Ms. Jones, "how far is the expected distance we have to walk before we reach the first station?"

"You mean," said Dr. Smith, "that if we repeated this event many, many times, what in the long run would be the average distance we would have to travel?"

"Precisely," said Ms. Jones. "Of course we have to include in each repetition the initial random selection of two points for the two supply stations. I figure it like this. The second station can be any distance from the first, from zero to the length of the crater's circumference. So the average distance will be half the circumference. Now we had the same chance of landing in one of those semicircles as the other. In either case our average distance from the two stations would be half a semicircle. Therefore the expected distance to the nearest station must be one-fourth of the circumference."

"Sounds plausible," said Dr. Smith, whose Ph.D. was

in statistics, "but that's *not* the right way to go about it. The expected distance is . . ."

What correct distance did Dr. Smith give, expressed as a fraction of the crater's circumference?

13

PINK, BLUE, AND GREEN

PUZZLE

The humanoids who live on a small planet in the Milky Way, not far from our solar system, are divided into three races with skins that are pink, blue, and green. We'll call them the pinks, blues, and greens. Like earthlings, they are bilaterally symmetric, each with two legs, two arms, and one head with eyes, ears, nose, and mouth.

One afternoon three college professors, with three different skin colors, were having lunch together in the faculty dining hall.

"Isn't it amazing," said Professor Pink, "that our three last names are Pink, Blue, and Green, yet not one of us has a name that matches his skin?"

"It is indeed a remarkable coincidence," agreed one of the others as he stirred his drink with a blue hand.

"The hall is crowded today," observed Professor Green, "but there seem to be very few greens in the room."

Pink glanced around. "Yes," he said. "There are more than three greens having lunch but certainly not more than a flink." ("Flink" is the local word for a dozen.)

"This gives me an idea for a brainteaser," said Professor Blue, the only mathematician in the group. "Perhaps Gardner can use it for his regular feature in *Isaac Asimov's Science Fiction Magazine*."

"How does it go?" asked Pink.

"Like this," said Blue. "I count exactly 80 pink arms on the persons sitting here, and half as many blue arms. If you add the number of pink persons to the number of blue persons, then add the number of eyes on all the greens, you get a grand total of 81. I wonder if Gardner's readers can deduce how many greens are having lunch?"

Green thought about the problem, then began to chuckle. "Excellent, Blue," he said. "You must radio it to Gardner as soon as we finish eating. Of course you should add that the count of legs, arms, and eyes includes us three, and that no one in the room is missing an arm, leg, or eye."

How many greens are having lunch?

14

THE THREE ROBOTS OF PROFESSOR TINKER

Professor Lyman Frank Tinker, head of the Artificial Intelligence Laboratory at Stanford University, was the twenty-first century's top designer of robots. One afternoon, for a seminar test, he brought to the classroom three female robots, all young, attractive, unclothed, and absolutely identical in appearance. He seated them on three chairs in front of the class.

"One of these girls," said Professor Tinker, "is programmed always to speak truly. Another is programmed always to lie. The third is programmed so she sometimes speaks truly, sometimes falsely. The decision is made by an internal randomizer. Your problem is this: In how few questions can you identify the truther, the liar, and the sometimer?"

The second smartest student in the class asked the following three questions:

1. To the lady on the left he said, "Who sits next to you?" The robot answered: "The truther."

2. To the center lady he said, "Who are you?" The robot answered: "The sometimer."

3. To the lady on the right he said, "Who sits next to you?" The robot answered: "The liar."

From the three answers the student correctly identified all three robots. How did he do it?

15

PUZZLE

HOW BAGSON BAGGED A BOARD GAME

Sidney Bagson was the twenty-fifth century's world expert on ancient mathematical games. He was delighted when a French archeologist, on a dig in what had once been New Jersey, found an artifact that seemed to be the board of an unknown twentieth-century game. The board's pattern is shown below. Somehow the artifact had survived the terrible world war of the early twenty-first century that had virtually obliterated North America.

Bagson had never seen such a board before. Try as he would, he was unable to deduce any reasonable playing rules. There was, however, a way to solve the mystery. The Weizmann Institute of Science, in Rehovot, Israel, where Bagson was a mathematician, owned a machine that enabled one to travel back in time and view past events. The machine and the person inside could not interact with the events being viewed. It had long been established that such interactions were impossible because they led to logical contradictions of the kind so thoroughly explored in primitive science fiction.

The Weizmann Institute gave Bagson permission to transport its machine to Secaucus, New Jersey, where the artifact had been unearthed. Extremely precise methods of quark dating established that the board had been made in the fall of 1987.

Bagson climbed into the machine, adjusted the dials, and soon found himself watching a boy and girl playing the game. They were, of course, unaware of his presence. After observing a few dozen games it was easy to deduce the rules:

1. The game begins with a red counter on spot X, a blue counter on spot Y. Players choose colors, then alternate turns in moving their counter forward along the zigzag line.

2. On each move a player must advance his counter 1, 2, or 3 spots along the line. No jumps are allowed.

3. When the pieces meet—that is, when they are adjacent

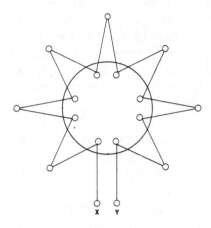

and no more moves are possible—the player whose piece is *inside* the circle wins.

Bagson at once recognized this as what mathematicians call a "nimlike" game. It cannot end in a draw because when the counters meet, one piece must be inside the circle, the other outside. It follows that the first or second player has a sure win if he or she plays correctly.

Which player can always win, and what strategy must the winner follow?

16

THE SHOP ON BEDFORD STREET

PUZZLE

I was ambling along Bedford Street in New York City's Greenwich Village, searching for the hidden entrance to Chumley's restaurant, when I passed a curious little shop that I had never noticed before. It was about two yards wide, even smaller than the narrow brownstone house on Bedford where Edna St. Vincent Millay burned her candle at both ends. A crudely lettered sign in the shop's unwashed window said nothing more than "Old SF and Fantasy Magazines."

I pushed open a decaying door. Behind a cluttered desk a gnomelike old man was snoring uncouthly.

"Do you have," I said loudly, "any copies of *Amazing Stories* prior to 1950?"

Two watery blue eyes opened slowly. The old man glowered at me while he picked his left ear with a mechanical pencil. "Of course."

He stood up wearily and climbed a creaky stepladder to take from a high shelf a stack of *Amazing Stories*. They seemed in astonishingly good condition, all dating from 1926 through 1949 and no two alike.

"The latest issue costs a dollar," the old man said. "The next costs three dollars, the next five, and so on in consecutive odd numbers. It's not a complete run. Lots of issues are missing. But you have to buy them all."

I counted the magazines and scribbled a calculation on the back of an envelope.

"I can't afford it."

"In that case," said the gnome, "I'll let you figure the price differently. You can divide the stack into two parts in any way you like, and pay for each according to the same system—a dollar for one, three dollars for two, five for three, and so on. I'll take a personal check, but you have to raise the amount to a multiple of a hundred dollars."

I doubt if the old man knew who I was, but in any case his crazy pricing scheme appealed to my fondness for number puzzles. I divided the stack so the total cost was as low as possible. To this price I added enough to bring it up to a multiple of a hundred, then I wrote a check. After looking over my driver's license and all my charge cards the gnome tied the dusty magazines into bundles, and I carried them to my car parked around the corner on Grove Street.

During dinner with my wife at Chumley's—she was annoyed by my being late—I told her about the bizarre bargain.

"Some bargain!" she said. "How much was the gratuity? I mean, how much did you add to the price to bring it up to a multiple of a hundred?"

"Let me put it this way," I said. "For one magazine, and only one, I paid exactly five times as much in dollars as your age."

"If you think I'm going to try to figure *that* out," she said, "you're nuttier than usual."

How many dollars did the old man get as a gratuity?

17

TANYA TACKLES TOPOLOGY

PUZZLE

The *Bagel*, a spaceship so-named because it was shaped like a bagel, was on its way back from the mission mentioned in puzzle 7.

"What's topology, Father?" asked Tanya, the ten-year-old daughter of Ronald Couth, head of the *Bagel*'s computer science crew.

"Roughly, it's the study of the properties of a structure that remain the same when the structure is distorted in a continuous way," said Couth. "But let me give you an old puzzle that will help you understand."

Couth found a piece of chalk, and on the bulkhead (wall) of the spaceship sketched the following pattern:

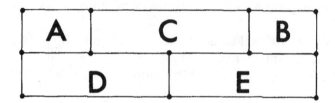

"The problem," said Couth, "is to draw one continuous line, starting anywhere you like and ending where you like, that will cut once and only once across each of the sixteen line segments in this figure. For example, you might draw a line like this." Couth drew a line that twisted around as shown below:

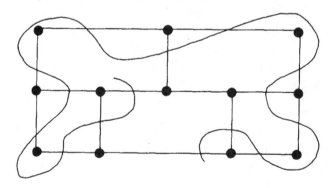

Tanya stamped her foot. "Now you've solved it!"

"Oh, no," said her father. "Look more carefully. You'll see that my line missed one segment."

"The reason this is a topological puzzle is that if you imagine the figure drawn on a sheet of rubber, and the rubber stretched any way you like, the puzzle remains unchanged. For instance, we could distort the figure like so:

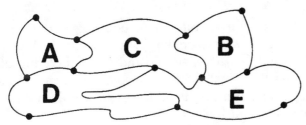

"Since the figure can be stretched and twisted, the question of whether the puzzle can be solved or not is topological. Actually there isn't any way to solve it unless you cheat by drawing your line through one of the vertexes or along one of the line segments."

"If it's impossible," said Tanya, "why should I waste time on it?"

"Because," said her father, "I want to see if you can prove it can't be solved. If you succeed, you'll learn something about how to prove certain theorems in elementary topology."

Can you prove that the puzzle has no solution?

18

THE EXPLOSION OF BLAB- BAGE'S ORACLE

PUZZLE

Professor Charles Blabbage, England's top expert on artificial intelligence, finally completed his construction of ORACLE, an acronym for Omniscient Rational Advance Calculator of Local Events. The computer was so powerful that it could (Blabbage maintained) predict with 100 percent accuracy any event in the laboratory within a period of one hour and inside a radius of ten meters from the computer's console.

This is how it operated. One could describe to ORACLE any event that would or would not occur during the next hour and within the specified radius. If the computer predicted that the event would take place it turned on a green light for "yes." If it predicted the event would not take place it turned on a red light for "no."

It was necessary, Professor Blabbage made clear, that the two lights be concealed until the hour was up. Otherwise anyone could easily render a prediction wrong by doing something to falsify it. For example, suppose the computer predicted "yes" to: "A cockroach will crawl across the west wall of the lab." If someone saw the green light he or she could stand guard by the wall to make sure the event did not occur.

Blabbage's assistant was Dr. Ada Loveface, an attractive young redhead with a doctorate in modern logic and set theory. On the day before Blabbage was to demonstrate ORACLE's powers for a group of distinguished visiting computer scientists, military moguls, and government officials, Dr. Loveface approached him and said:

"I regret having to tell you this, Professor, but I've just proved that ORACLE can't possibly succeed in all cases. I can describe an event that will or will not take place in the lab, within the hour and inside the ten-meter radius, of such a nature that the computer will find it logically impossible to predict whether it will or won't happen."

Blabbage refused to believe Ada until she told him

what the event was. Her remarks were so shattering that he collapsed in a faint and had to be taken to a hospital.

What event did Dr. Loveface think of?

19

PUZZLE

DRACULA MAKES A MARTINI

"It's cocktail time, my love," said Count Dracula to his wife. "Shall it be the usual?"

"The usual," said Mrs. Dracula.

The count took from his liquor cabinet a bottle containing one quart of vodka and a smaller bottle containing one pint of human blood. He poured a small quantity of blood into the vodka, shook the bottle vigorously, then poured exactly the same amount back into the bottle of blood. Hence at the finish there was again a quart of liquid in the large bottle and a pint in the small bottle.

Mrs. Dracula was sitting with her back to her husband, but she was watching him in a mirror on the living-room wall. The count was following the standard Transylvanian procedure for making a vampire martini.

Assume that when vodka and human blood are mixed, neither alters in volume. After the two operations just described, is there more vodka in the pint of blood than there is blood in the quart of vodka, or less, or are the two amounts the same?

You may have come across this puzzle before in the form of identical glasses, one filled with water, the other with wine. In this variant, however, the contents of the two containers are *not* alike, nor are we told the amount of liquid that is transferred back and forth.

20

THE ERASING OF PHILBERT THE FUDGER

By the mid-twenty-third century capital punishment has been replaced throughout most of the civilized world by a punishment called "erasure." The criminal's head is placed inside an electronic machine called the "oblivion box." It takes only a few minutes to expunge from the brain all memories of events experienced after the first six months of life. This, of course, returns the criminal to babyhood. It has long been established that no one is born with criminal tendencies—all are acquired by experience. Over a period of years the erased "baby" slowly develops into a new adult. Because erasure turns a person into a different personality, with no memory of his or her former self, the punishment is feared almost as much as execution.

Another radical change in the administration of justice is the replacement of all judges, and some lawyers, by robots. Laws have become so numerous and complicated that only computers can remember all the details. Robot judges are carefully programmed to make only wise and logical decisions. It is impossible for a robot judge to lie. If a circuit in his brain malfunctions, and he makes any statement that is false, his pronouncements are declared null and void and a new trial is scheduled.

One of the most heinous crimes in the twenty-third century, on a par with murder and rape, is the crime of "fudging." This means a falsification of data in a scientific experiment. Such an enormous respect has developed for the sanctity of scientific method that anyone declared guilty of fudging is automatically sentenced to erasure.

Philbert X1729B was arrested for having fudged the data in a tooth-decay experiment he had supervised at the Oral Roberts Dental Research Laboratory, in Tulsa, Oklahoma. Philbert could not afford a robot lawyer. His human lawyer, who was not very skillful, lost the case. At Philbert's sentencing the robot judge said:

"You will be erased at 3 P.M. on one of the six days of

next week, starting with Monday. You'll be informed of the day at 10 A.M. on the day of the erasure."

"But judge," said Philbert, "can't you tell me the day now?"

"No. The date has not yet been determined. I can assure you, however, that it will be Monday, Tuesday, Wednesday, Thursday, Friday, or Saturday of next week. You won't know what day it is until we inform you on the morning of erasure day."

"Thank you, your honor."

It was the last sentencing of the day, so the judge pressed a button under his left armpit to turn himself off until the court opened the next morning.

Sitting in his cell, Philbert began to think about what the judge had said. Suddenly he leaped to his feet with a yelp of joy. There was no way he could be erased without making the judge out to be a liar! This would guarantee him a new trial. Maybe his wife and friends would be able to raise enough money for a good robot lawyer.

What was Philbert's reasoning?

21

OULIPO WORDPLAY

There is a French group of eminent mathematicians and writers who call themselves the Oulipo—a name that derives from *Ouvroir de Littérature Potentielle*, or Workshop of Potential Literature. They are dedicated to every variety of wordplay, especially the devising of "algorithms" for transforming poems and prose passages in such a way that the result is almost, but not entirely, nonsense.

One famous Oulipo algorithm operates as follows: Take the first sentence of a short story and number its nouns from 1 to *n*. Unhyphenated two-word nouns, such as "Volume 5," are treated as single nouns. Now go to the end of the story and number the nouns from 1 to *n*, taking them in *backward* order from the end. Return to the story's first sentence. For its first noun substitute the story's last noun, for its second noun substitute the second noun from the end, and so on until the *n*th noun of the opening sentence has been replaced by the *n*th noun from the end.

The result usually makes a strange kind of sense that conveys something of the story's mood and the author's style. One may take liberties with capitalization, and also alter singular nouns to plural (or vice versa) when this is necessary to make a sentence grammatically correct.

The six sentences below resulted when I applied this crazy algorithm to the opening sentences of six well-known science fiction stories by eminent writers:

1. Three hundred stars and more from night, one hundred from the sunset of glow, in the wildest king of Blind Valley, there lies that mysterious mountain thing, cut off from the sky of vastness, the darkness of the purple.

2. Actions are considered crazy anywhere in the man.

3. By the day he reached the morning of the little evening, even the gardens of his walls were drained.

4. "This is a slightly unusual star," said Dr. Fuss, with what he hoped was commendable everything.

PUZZLE 21

5. Parts took time of the month off the Methuen and opened it to chemistry.

6. Night, sun of glow, thrust out a belligerent lower brightness and glared at the young glow in a hot city.

Can you identify the authors and story titles?

22

HOW CROCK AND WITSON CRACKED A CODE

In the year 2019 a remarkable bacteriophage (a virus that infects bacteria) was discovered by Dr. Frank Crock, a bacteriologist at Harvard University.

"I can't believe it," said Crock to his associate, Dr. James Hugh Witson. "The DNA message in this virus is the simplest and strangest I've ever seen. It just keeps repeating the same sequence of a dozen triplets."

All DNA information, along the double helix, is given by a sequence of four bases: adenine, cytosine, guanine, and thymine. They are grouped as three-letter "words," using the four-letter "alphabet" of the initials A,C,G,T. The new phage contained a DNA "message" that consisted of twenty repetitions of the following thirty-six-letter sequence:

GTT ATG TCC CTC TCA CTC TCC CTC ACG CTC TGG AGA

"It certainly looks artificial," said Witson. "Could it be that the phage was sent here by extraterrestrials?"

"I wouldn't rule it out," said Crock. "Back in 1979 two Tokyo scientists suggested that some of our viruses might be artificial and sent here as coded communications from another planet. It would be an efficient way to communicate because the virus would replicate rapidly until it covered the earth. If I remember right, the two Japanese scientists actually searched the DNA sequences of many simple viruses, looking for some sort of communication. They failed to find any. But this virus couldn't possibly have evolved here. It *must* be artificial."

Crock and Witson began an intensive study of the thirty-six-letter sequence to see if they could find anything that resembled a communication from an alien intelligence. It took only a short time to discover that one of the letters marked a familiar sequence of integers that could not have been a coincidence. What sequence did they find?

23

PUZZLE

TITAN'S TITANIC SYMBOL

Larc Snaag, captain of the *Bagel*, maneuvered the mammoth spaceship closer and closer to Titan. It was earth's first manned probe of this giant satellite of Saturn, a moon larger than Mercury and almost the size of Mars.

"Great Isaac!" shouted Snaag's executive officer. "Look!"

A geometrical figure, made with glowing green lines, could be seen distinctly through Titan's swirling atmosphere. It was an equilateral triangle inscribed in a circle which in turn was inscribed in a larger equilateral triangle.

"It's ... it's titanic!" bellowed the captain. "What in Scitheration do you suppose it means?"

"Could be a religious symbol," said the exec. "Or

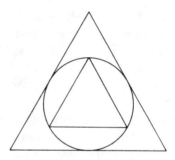

maybe it's just a way of letting us aliens know that a brainy civilization is flourishing under those bloody clouds."

During chow, the mysterious symbol was, naturally, the topic of excited conversation among the *Bagel*'s crew. Someone asked Ronald Couth, the ship's top mathematician, to calculate the ratio between the areas of the two triangles.

"That shouldn't be hard," said Couth as he started to draw on a napkin. "We let the circle's radius be 1, then construct a right triangle like so. Now we can apply the Pythagorean theorem, and—"

"And," said Couth's bright ten-year-old daughter, Tanya, who had been watching with a grin, "you find that the big triangle is exactly four times the small one."

"Maybe," said Couth as he scribbled a quadratic equation. Five minutes later he turned to his daughter in astonishment. "You're right! But how did you solve it so fast?"

"Easy," said Tanya. "I just noticed that . . ."

What "aha!" insight did Tanya have that produced an immediate solution?

24

PROFESSOR CRACKER'S ANTITELE-PHONE

PUZZLE

After the explosion of Charles Blabbage's prediction machine (see puzzle 18), his assistant, Ada Loveface, went to work for Alexander Graham Cracker, a famous astrophysicist at Barkback College in London.

Professor Cracker's project was to design a machine that could, at least in theory, send signals through interstellar space at speeds faster than light. Back in the twentieth century, physicist Gerald Feinberg and others had found that relativity theory permits the existence of particles that always go faster than light. Feinberg called them "tachyons" after a Greek word for "swift."

Just as ordinary particles ("tardyons") can never be accelerated to the speed of light, so tachyons can never be slowed down to the speed of light. Since tachyons are always moving, they have no rest mass. This allowed Feinberg to represent their rest masses by imaginary numbers. After six months of intensive research, Professor Cracker finally designed what he called a "tachyonic antitelephone." Although tachyons had not yet been proved to exist, *if* they existed Cracker's antitelephone could modulate a beam of tachyons in such a way that signals could be carried by the beam.

"Because tachyons move faster than light," said Dr. Loveface, "doesn't that mean they travel backward in time?"

"Of course," replied Professor Cracker. "Einstein's equations guarantee it. That's what's so marvelous about my antitelephone. We can send a message to intelligent aliens in the Andromeda galaxy at such a speed that it gets there several days before we send it!"

"In that case," said Dr. Loveface, "your antitelephone won't work."

Dr. Loveface then outlined a logical proof of her statement that was so ironclad that Professor Cracker abandoned his project at once. What sort of proof did she give?

25

VACATION ON THE MOON

Edgar D. Twitchell, a New Jersey plumber, was on his way to the moon for a three-week holiday. The rocket ship was too small to generate artificial gravity by spinning, so Twitchell had the strange sensation of feeling his weight steadily diminish as the ship sped toward its destination. When it reached the spot where earth's stronger gravity field was exactly balanced by the moon's weaker field, zero g prevailed inside the ship. All passengers were kept fastened to their seats, but Twitchell enjoyed the floating feeling nonetheless as he twiddled his thumbs and contentedly puffed a cigar.

Many hours later the ship slowly settled next to one of the huge domes that house the U.S. moon colony, its descent cushioned by rocket brakes. Through the thick glass window by his seat Twitchell caught his first glimpse of the spectacular lunar landscape. Several large sea gulls, with tiny oxygen tanks strapped to their backs, were flying near the dome. Above the dome an American flag fluttered in the breeze.

Although it was daylight, the sky was inky black and splattered with twinkling stars. Low on the horizon a rising "New Earth" showed a thin bluish crescent of light with several faint stars shining between the crescent's arms. As Twitchell later learned, the moon makes one rotation during each revolution around the earth. Because a rotation takes about twenty-eight days, it takes the earth about fourteen days to rise and set on the moon.

On the sixth day of his vacation, Twitchell was allowed to put on a space suit and hike around the crater in which the dome had been built. After bounding along for a while he came upon a group of children, in pink space suits, playing with boomerangs. One girl tossed a boomerang that made a wide circle and Twitchell had to duck as it whirled past his helmet. Behind him he heard it thud against a large boulder. He turned to look, but the curved stick had fallen into the rock's ebony shadow where it instantly seemed to vanish. Since there is no atmospheric scattering of light on the moon, objects cannot be seen in shadows without a flashlight.

The sun was low in the sky when Twitchell began his walk. Now it was sinking out of sight. The "terminator," that sharp line separating the lunar day from night, was gliding across the gray terrain toward the brightly lit dome at a speed of about 40 miles an hour—much too fast for Twitchell to keep up with it by vigorous hopping. Overhead a meteor left a fiery trail as it fell to the moon's surface.

Twitchell was so exhausted when he returned to his quarters that he fell asleep on his bed, fully clothed, and did not awake until the rising sun flooded his room with brilliant sunlight.

How many scientific mistakes can you find in the above narrative?

After the crew of the spaceship *Bagel* observed a geometrical symbol glowing below the orange clouds of Titan (see puzzle 23), it was obvious that intelligent life flourished on Saturn's largest moon.

"If the Titans went to all that trouble of constructing such a mammoth symbol," said the *Bagel*'s captain, Larc Snaag, "it must be because they want others in the solar system to know they're there. If so, surely they must also be broadcasting a radio message to outsiders."

Frank Blake, the crew's chief radioman, was at once ordered to make a thorough search of all radio frequencies. A few hours later he reported, in great excitement, that a coded message was coming through loud and clear.

It's a series of beeps," said Blake, "separated by pauses. They're sending a curious number sequence."

"The primes?" asked Snaag. "Or maybe pi or the square root of two?"

"No, nothing that simple." Blake handed Snaag a sheet on which the following sequence was written:

1   3   7   12   18   26   35   45   56   69
83   98....

"The sequence goes up to a hundred numbers," said Blake, "then it's repeated over and over again. VOZ [the ship's computer] is studying it now."

A short time later Ronald Couth, the *Bagel*'s computer officer, burst into the captain's quarters. "It's beautiful! VOZ says the sequence was discovered in the 1970s by Douglas R. Hofstadter, who called them 'weird numbers.' He gave the sequence on page 73 of his classic 1979 book, *Gödel, Escher, Bach: An Eternal Golden Braid.*"

"A marvelous book," said Snaag. "I read it when I was in college. But I don't remember the sequence. How's it formed?"

Couth jotted down the first six numbers, then under each pair of numbers he put the pair's difference:

1	3	7	12	18	26 ...
2	4	5	6	8 ...	

"The sequence," said Couth, "and its first row of differences contain all the counting numbers, each number appearing just once. It's unique if we assume that the numbers in both rows, as well as the two starting numbers of each row, are in increasing order. The construction method is easy. Begin with 1. The next number can't be 2 because that would duplicate 2 in the row below, so it must be 3. The next number can't be 4, 5, or 6 because that would put a 1, 2, or 3 in the row beneath, so it must be 7. And so on. It's obvious that the two rows catch all the positive integers, although the numbers at the top increase in size much faster than those beneath."

"I wonder," said Snaag, "if this can be generalized. Is there a sequence with first and second rows of differences such that the three sequences catch all the counting numbers, with no duplicates?"

"No," said Couth, "but it was clever of you to ask. VOZ reports that in 1980 Karl Fox, at Bell Laboratories in Columbus, Ohio, proved that there is no 'doubly-weird' sequence of numbers."

"Bell Labs?" said Snaag. "What's that?"

"They used to make a communication device," said Couth. "It was so uncouth that it transmitted by heavy cables that formed a gigantic web all over the United States."

Can you reconstruct Fox's proof?

27

LUCIFER AT LAS VEGAS

Ever since Satan's tumble from Heaven, his paranormal powers have varied directly with the strength of humanity's belief that he possessed such powers. As this belief faded, so did the Devil's ability to do devilish deeds. By the mid-twenty-first century, when nobody even believed he existed, the Devil's psychic powers had become so feeble that he could do little more than pronounce trivial curses that lasted no longer than twenty-four hours.

To avoid endless boredom in Hell, the Devil, disguised as a mortal, frequently visited the casinos of Las Vegas. It was hard to say which he enjoyed most, high rolling or the hookers. On this occasion he was playing the role of a tall oilman from Fort Worth.

"Care to make any side bets?" he asked a rotund man from Omaha who was standing near a roulette table.

"Depends on the bet."

"Naturally," said the Devil. "What I have in mind is this. Pick any triplet of blacks and reds, say red-red-black or black-red-black—any combination you like. Then I'll pick a different triplet. We'll agree on when to start, then we'll watch the spins to see which of our triplets shows up first as a run. If yours comes first, you win. If mine comes first, I win. We'll ignore any zero or double zero. I'll give you 5 to 4 odds—five of my ozmufs to your four. [An ozmuf is worth about 25 1980 U.S. dollars.] Each time we repeat the bet you can have the first choice of a triplet."

"Hmmm," said the fat man. "On every spin the probability is ½ for red, ½ for black. For any given triplet the probability is ½ times ½ times ½, or ⅛, so all triplets are equally likely to show. Neither of us has an advantage."

"Precisely," replied the Devil, smiling. "But I'm offering you better than even odds."

"Sounds like a hell of a good proposition," said the man from Omaha.

Whom does the bet favor, the man or the Devil? Does it make any difference which triplet the man picks each time?

28

OFF WE'RE GOING TO SHUTTLE

PUZZLE

You live in Mars, Pennsylvania, but you work at a cheese factory on the moon. Each Monday morning you helicop to Pittsburgh where you take an ion-powered shuttle ship to the moon, then you return to earth on the following Friday.

Except on holidays such as New Year's and Asimov's Birthday, a lunar shuttle ship takes off every even hour (12, 2, 4, 6, . . .), day and night, from the Pittsburgh launching station. Every even hour a ship leaves the moon for earth. In both directions the ships maintain a constant distance from one another, and their speeds are adjusted so that they arrive at each destination on the odd hours (1, 3, 5, 7, . . .). The illustration shows a number of the ships leaving earth, and an equal number arriving. One day on your way to the moon you see the red port light of an earth-bound ship streak

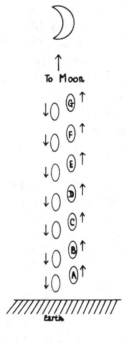

across the ebony sky like a fiery meteor through earth's atmosphere.

"I've been counting the incoming ships," says the young lady seated next to you and by the window. "That's the seventh ship we've passed since we blasted off."

On which ship (A through G) are you riding? Try to answer this in your head without modeling the problem by moving counters across a tabletop.

29

THE BACKWARD BANANA

Backward, turn backward, O time, in your flight.
Make me a child again just for tonight!
—ELIZABETH AKERS ALLEN

Professor N. A. Gilligan and his assistants, Bianca Zacnaib and Duane Renaud, had been working for years on a device they hoped could reverse time inside a small region of space. Their method is much too technical to put in laymen's language, but essentially it involves a reversal of the spin of Penrose twistors—mathematical structures that underlie quarks. Twistors had been proposed in the mid-twentieth century by the British mathematical physicist Roger Penrose, and their existence was confirmed in 1991 by a series of ingenious experiments.

Gilligan's device was slightly larger than a washing machine. A compartment at the top, supercooled and surrounded by a powerful Penrose force field, was designed to hold any physical object that the experimenters intended to time-reverse. On the side of the machine were twenty levers, their positions numbered 1 through 20. Pulling up on a lever closed a position, pushing down opened it.

"At last we are ready for a test," said Gilligan, his eyes gleaming. "Let's first use a small organic object, say a lemon. If we succeed, we'll try to time-reverse a watermelon. Then maybe a mouse."

"No lemons, no melon," said Bianca, "but we do have a banana in the refrigerator."

"Banana it is," said Gilligan.

Bianca fetched the ripe yellow banana. She carefully placed it inside the compartment and closed the lid.

"Are all positions closed?" asked Gilligan.

"No, it is open on one position—position 2," said Duane. "Shall I close 2?"

"Not yet," said Gilligan, walking around the machine to inspect the row of levers. He pushed down lever 7, adjusted several dials, then pressed a button that turned on the machine. It began to hum.

Gilligan hooked a finger under lever 7 while Duane kept a finger below lever 2. "Pull up if I pull up," said Gilligan.

Gilligan waited a few minutes before he slowly raised his lever. Duane did the same. "Bianca, as I move these levers," said Duane, "I'm so excited my hand is shaking."

Bianca raised the lid for a quick peek. The banana had already turned green.

The hum grew steadily louder, then suddenly there was an explosive sound, like a tiny thunderclap, inside the compartment. When Bianca opened it again, the banana was gone.

"By Albert, we've done it!" shouted Gilligan.

The three physicists broke open a bottle of wine, drank several toasts, sang a chorus of *Yes, We Have No Bananas,* then went to Gilligan's office to prepare their report on the great experiment.

Did you notice that in the above episode the names of all three scientists are palindromes? That is, the order of letters in each name is the same when the letters are taken in the reverse direction.

Concealed in the text are three other palindromic word sequences that spell the same backward. One contains just four words, one contains six, and one contains seven.

30

PUZZLE

THE QUEER
STORY OF
GARDNER'S
MAGAZINE

One of the least known of all science fiction tales by H. G. Wells is "The Queer Story of Brownlow's Newspaper." It appeared in the *Ladies' Home Journal*, April 1932, and has never been reprinted in any book.

Because of some strange time warp, a newspaper dated November 10, 1971, is delivered to Brownlow forty years ahead of time. The story is mainly a description of what was in the paper. Wells made a few lucky hits (such as lower birth rates, emphasis on psychological motivation in fiction, attempts to utilize heat below the earth's surface, wider coverage of science news), but these are balanced by Wells's misses (simplified English spelling, world government, no financial pages, thirteen-month calendar, the gorilla has become extinct, and others).

Last January I had an experience similar to Brownlow's. A copy of what I assumed would be the January 1980 issue of *Scientific American* arrived in the mail. Incredibly, it was dated January 2556! The printing was peculiar, and the language hard to understand, but the illustrations were spectacular—all in full color and three-dimensional. Many of them moved when you tilted the page.

I turned at once to the Mathematical Games column. Written by someone using the obvious pseudonym of Nitram Rendrag, the column was devoted entirely to number puzzles involving the date of the new year. Here is a selection of six I could understand.

1. Scramble the digits of 2556 any way you like and enter the number in your calculator. Multiply it by any digit, add 100, and divide by 3. The quotient always has a fraction of 1/3. Why?

2. Inside the nine squares below, put the digits 1 through 9, using each digit just once. Arrange them so the rows add to the numbers shown on the right, and the three three-digit numbers add to 2556. The pattern is unique.

$$\square + \square + \square = 18$$
$$\square + \square + \square = 15$$
$$\square + \square + \square = 12$$

2   5   5   6

3. Between any adjacent pair of digits in the sequence
$$1\ 2\ 3\ 4\ 5\ 6\ 7\ 8\ 9$$
insert either a plus sign or a minus sign or nothing at all. Digits without signs between them form larger numbers. For example: $123 - 45 - 67 + 89 = 100$. This is the only way, said Rendrag, to obtain a sum of 100 with as few as three signs.

Your task is to use as many plus or minus signs as you like to make the sum 556. There is only one solution.

4. Using just three plus or minus signs make the sum 56. This also has just one answer.

5. Form an expression for 56 using the digit 4 no more than three times, plus any of the following symbols: $+, -, \times, \div, !$ (the factorial sign), the radical sign, the decimal point, and parentheses. Exponents may be shown, but repeating decimal fractions cannot be indicated by a dot above a number. A permitted symbol may be used as often as you wish.

6. Circle any number in the matrix on the next page, then cross out its row and column. Circle any number not crossed out, and again eliminate its row and column. Repeat this two more times. Circle the only remaining number. Add the five circled numbers. The sum will be 56. Why does this seeming miracle always work?

10	12	13	9	11
9	11	12	8	10
13	15	16	12	14
11	13	14	10	12
8	10	11	7	9

As customary, Rendrag withheld his answers until the following month.

31

BLABBAGE'S DECISION PARADOX

After twenty years of work, Professor Charles Blabbage, whom we met in puzzle 18, finally perfected his notorious decision prediction machine. Working details are too technical to explain, but essentially the device scans a human brain with three mutually perpendicular neutrino beams. Information on all electrical activity inside the skull is then analyzed by a powerful bubble computer. When any person is faced with a decision between two mutually exclusive courses of action, the machine can predict with amazing accuracy how he or she will decide.

For several months Professor Blabbage had been working with an amiable subject named Robert Zonick, obtaining an average success of 98 percent for all predictions.

"I have a new and curious test for you today, Bob," said the professor. "Observe that there are two boxes here on the table—one transparent, one opaque."

Bob nodded as he took his usual seat beside the table. Blabbage moved the three neutrino guns to within a few centimeters of Bob's forehead, his left temple, and the crown of his head.

"As you can see," Blabbage continued, "there is a hundred-dollar bill inside the transparent box."

"And the opaque box?" Bob asked, pointing his finger.

"At the moment it's empty," said the professor. "But let me explain." He glanced at his wristwatch. "One hour from now I'll ask you to make one of two choices. Either choose the opaque box only, or choose both boxes. If my machine predicts that you will take the opaque box, I'll put inside it a cashier's check for a million dollars. The money will be yours."

"Marvelous!" grinned Bob. "This test I like!"

"However, if my machine predicts you will take both boxes, I'll put *nothing* in the opaque box. Of course, you'll be certain then to get the hundred dollars."

Professor Blabbage pushed a button and the machine buzzed for a few seconds. Then he picked up the opaque box and left the room. A half hour later he returned to put the box on the table beside the transparent one.

"The computer has determined how you'll probably decide," he said. "But think it over carefully. You have twenty minutes to make up your mind. Of course you must not touch either box until you make your choice. Everything is being videotaped. If the opaque box is empty now, it will be empty then. If it has the check inside it now, it will be there when you open it. Good luck, my friend."

After Blabbage left, Bob stared at the boxes for several minutes. "I've been tested a hundred times with this infernal machine," he said to himself, "and it was almost always right. So I should take only the opaque box. The odds are better than 9 to 1 that I'll get the big check. On the other hand . . . "

Bob suddenly realized that there was just as good an argument, maybe one even better, for taking *both* boxes! What is the argument?

32

NO VACANCY AT ALEPH-NULL INN

Our universe has an enormous but finite number of suns, and consequently a large but limited number of planets. Although the number of intelligent beings on these planets is much larger than the number of planets, it too is still a finite number.

However, an infinity of universes lie side by side in a higher spacetime just as the two-dimensional leaves of this book lie side by side in our three-dimensional world. Spinning at the center of the Milky Way galaxy is a black hole. An opening in the hole's singularity leads into the Black Tube, a tube that extends like a monstrous worm along the fourth coordinate of space and provides easy access to the infinite number of parallel universes. Inside the Black Tube is a lavish resort hotel known as the Aleph-Null Inn.

The inn is rather large. In fact it has an *endless* number of rooms. The rooms are numbered 1, 2, 3, 4, 5, . . . and so on to infinity. The hotel is a popular vacation spot for intelligent beings living in the infinity of parallel universes that are reachable through the Black Tube.

On one occasion, when all rooms of the Aleph-Null Inn were occupied, a creature crawled off a spaceship from Andromeda. He entered the inn and loudly demanded a room.

"Do you have a reservation?" asked the clerk, a female who vaguely resembled what on earth we call a kangaroo.

"I do not," rasped the creature, who looked like nothing on earth. "I didn't know I needed one. Don't you have an aleph-null number of rooms?"

"We do," said the clerk. "But at the moment every room is occupied."

The creature extracted a thousand-georg bill from his analog of what we call a wallet. "Maybe this will help you find a room for me."

The kangaroo, after glancing around the lobby to make sure no one was watching, quickly slipped the money into her pouch. "I think we can accommodate you," she whispered.

How did the clerk find a room for the creature from Andromeda without forcing any occupants to double up or to leave the inn?

33

TUBE THROUGH THE EARTH

In the twenty-third century an enormous gravity transport tube, with a diameter of 20 meters, was constructed straight along the earth's axis to join the metropolises of North Polaris and South Polaris. Through this tunnel cylindrical cars carrying both supplies and people were dropped from one city to the other. All friction was eliminated by maintaining a vacuum inside the tube, and by using magnetic fields to keep the cars away from the tube's side. The trip from pole to pole took only slightly longer than 42 minutes.

How many of the following questions about the transport tube can you answer?

1. As the car travels from North Polaris to the earth's center, does its velocity increase, decrease, or stay the same?

2. Does the car's *acceleration* increase, decrease, or remain the same?

3. If you are riding in a car and it stops halfway down to the earth's center, would you weigh less or more on a spring scale than on the earth's surface?

4. At what point during the trip would you experience zero gravity?

5. At what spot does the car reach maximum speed, and how fast is it going?

6. If a car fell down a similar tube through the center of the moon, would the time for a one-way trip be shorter or longer than 42 minutes?

7. A famous SF story was written about an attempt to dig a deep hole below the earth's crust. It turns out that the earth is a living organism, and when its epidermis is punctured the earth lets out a mighty yell of pain. What is the story's title and who wrote it?

34

PUZZLE

ROBOTS OF OZ

If by robot we mean a machine constructed to simulate a human being, there are three unusual robots that L. Frank Baum placed either in Oz or in one of the many magic realms just outside Oz. The Tin Woodman is not, of course, a robot but a former woodchopper whose human parts were gradually replaced by tin as they were chopped off by his enchanted ax. Here are three puzzles, each related to one of Baum's robots.

1. The Giant with the Hammer

In *Ozma of Oz* a giant made of cast iron guards a narrow path leading to where the evil Nome King lives in Ev. The giant neither thinks nor speaks. His sole task is to pound the path continually with an iron hammer about the size of a barrel.

Now for an easy brainteaser. Assume that it takes one second to raise the hammer, and one second to lower it with a loud bong on the road. How many seconds elapse between bong one and bong one hundred?

2. Tiktok

Tiktok also makes his first appearance in *Ozma of Oz*. As all Oz buffs know, he is a mechanical copper man made by Smith and Tinker, who also made the iron giant. But Tiktok is a much more complicated machine. He can think and speak as well as act. In fact, as Baum tells us, he "does everything but live."

Tiktok's actions, thoughts, and speech are each controlled by a separate windup mechanism that can run down independently of the others. When his thought machinery runs down, but not his talking machinery, he scrambles words.

One day, when Tiktok was reciting a well-known four-line poem by an American humorist, his thinking mechanism ran down but his speech mechanism somehow managed to rearrange the words to make a nonsense poem. Here is what Tiktok said:

> I hope, but I never can cow you.
> Rather I'd tell than be
> A one-to-one anyhow see-saw.
> Purple I never see.
>
> —by Lester Bustegg

Can you unscramble the words and reconstruct the original quatrain? I have taken liberties only with the punctuation and capitalizations. The author's name is an anagram of his real name.

3. Mr. Split

Mr. Split appears in Dot and Tot of Merryland, a delightful fantasy by Baum that has long been out of print. The book was illustrated by W. W. Denslow, illustrator of *The Wizard of Oz*. Merryland is near the northeast corner of Oz, just on the other side of the Deadly Desert that surrounds Oz.

The Sixth Valley of Merryland is inhabited by living windup toys. Mr. Split is a wooden robot whose job is to keep all the toys properly wound. To save time, he can unhook his red left side from his white right side. Each side can then hop about on one leg and wind up toys twice as fast as Mr. Split could manage when not split in half.

Unhooked, Mr. Left Split speaks only the left halves of words, and Mr. Right Split speaks only the right halves. Assume that if a word has an odd number of letters, the middle letter is omitted by either half of Mr. Split. Below are six familiar proverbs spoken by Mr. Right Split. Can you supply the full words?

> tch n me es ne
> e o ates s st

ing ne ers o ss
rd n e nd s th o n e sh
ok ore u ap
res o ce ke me

35

I was up late, working for hours on digital problems with a pocket calculator. Its light-emitting diodes displayed the digits in green, each formed by a subset of the seven separately wired bars of the pattern shown below:

THE DANCE OF THE JOLLY GREEN DIGITS

I finally went to bed at 3 A.M., and during the night had a strange dream. The ten digits, each as large as a person and glowing a ghostly green, came dancing into my bedroom. The group's leader was Zero, whose round face strikingly resembled that of the late Zero Mostel.

"We have the pleasure," Zero sang out in deep bass tones, "of entertaining you with some unusual digital curiosities. For our first number—the largest square that uses all of us except me. It's the square of 30,384."

The ten digits danced themselves into this formation:

923,187,456.

"If I join my jolly friends," said Zero, "the largest square is this." The digits danced around furiously, chuckling to themselves, then finally stood in a row:

9,814,072,356.

"It's the square of 99,066," Zero announced. "And that's a number that is the same upside down." When Zero snapped his fingers, seven digits left the row to leave 9, 0, and 6. Then an amazing thing happened. Suddenly the 9 and 6 each became twins to form the number 99,066. Then the nines and sixes traded places and they all stood on their heads to produce 99,066 again.

"Next," Zero called out, "is the *smallest* square without me." The nine digits formed 139,854,276, the square of 11,826. "And the smallest *with* me." The ten digits danced to new positions to make 1,026,753,849, the square of 32,043.

"Here's something different," said Zero. The nine non-zero digits danced to positions in which two of them formed 28, three formed 157, and the remaining four formed the product 4,396.

"That's the lowest product for a twin and a triplet," boomed Zero. "The highest is this." He clapped his hands and the digits danced to $48 \times 159 = 7,632$.

"Of course if I join them we get different solutions," said Zero. The digits first formed the solution with the smallest product, $39 \times 402 = 15,678$, then the solution with the largest product, $63 \times 927 = 58,401$.

"Now," said Zero, "for a truly remarkable number." First the digits formed 87,021 and 94,356. Then they separated to reform the product 8,210,953,476. Zero claimed it was the largest number of ten different digits that is the product of two five-digit numbers that contain all ten digits.

Zero made a low bow. "I know you need puzzles for Isaac's science fiction magazine," he smiled. "Here's a funny oldie." The digits roared with laughter as they soft-shoe shuffled into the formation

8,549,176,320.

"What," asked Zero, "is so remarkable about this number?"

36

THE *BAGEL* HEADS HOME

When the spaceship *Bagel* returned from its mission to Titan (see puzzle 23), it went first to the U.S. moon base for repairs. Two weeks later it was on its way to the earth from the moon.

Ronald Couth, who headed the *Bagel*'s computer science crew, was playing a game of go with VOZ, the ship's computer, when his daughter Tanya, now twelve, entered the computer shack. "I just noticed something unusual," the girl said. "First I looked at the earth through a front window. Then I went to the back of the ship and looked at the moon. They look exactly the same size!"

Colonel Couth smiled. "Of course you know there's just one spot along the way where that happens, and locating the spot on a chart is a good exercise in geometry. To simplify the problem, let's round off all the relevant dimensions. Assume the distance from the moon's center to the earth's center is 240,000 miles; the earth's diameter is 8,000 miles; and the moon's diameter is 2,000 miles. Do you think you can figure out how far we are now from the moon's center?"

Tanya, who loved geometry problems, had no trouble with this one.

ANSWERS

1

Dr. Ziege could have started from any spot on a circle about 11.59 myriameters from Capra's *south* pole. Driving 10 myriameters south would take her to a spot $5/\pi$ myriameters from the pole. Now if she drives 10 myriameters east she will complete one full circle around the pole. Continuing 10 myriameters north returns her to where she started.

The rescue party found Dr. Ziege and her companions where Felix had predicted, and in time to save their lives. On the way back to earth Hilda suddenly realized there was a *third* spot on Capra from which Dr. Ziege could have started! Can you identify the third spot?

2

Let 1A stand for the insides of the first pair of gloves, 1B for the outsides. Let 2A stand for the insides of the second pair, 2B for the outsides.

Dr. Xenophon wears *both* pairs, the second on top of the first. Sides 1A and 2B may become contaminated. Sides 1B and 2A remain sterile. Dr. Ypsilanti wears the second pair, with sterile sides 2A touching his hands. Dr. Zeno turns the first pair inside out before putting them on. Sterile sides 1B will then be touching his hands.

After Dr. Zeno finished operating, his nurse, Ms. Frisbie, was furious. "You boneheads ought to be ashamed! You protected yourselves, but forgot about poor Ms. Hooker. If Dr. Xenophon has the flu, Ms. Hooker could catch it from the gloves you and Dr. Ypsilanti wore."

"Are you suggesting, Ms. Frisbie," asked Dr. Zeno, "that we could have prevented that?"

"That's *exactly* what I'm suggesting."

Then, to Dr. Zeno's amazement, Ms. Frisbie explained how they could have followed another procedure that would have eliminated not only the possibility of the surgeons catching the Barsoomian flu from one another or from

Ms. Hooker, but also the possibility of Ms. Hooker catching it from the surgeons. What was Ms. Frisbie's explanation?

3

Mathematicians call this a Diophantine problem. A Diophantine equation is an algebraic equation to be solved with integral values. In this case the equation is:

$$\frac{x(x + 1)(x + 2)}{6} = \frac{y(y + 1)}{2}$$

The left expression defines tetrahedral numbers, the right expression defines triangular numbers. Values for x and y must be positive integers. We already know two solutions:

$x = 1, y = 1$.
$x = 3, y = 4$.

The two solutions give values of 1 and 10 for the number of balls. The next solution is:

$x = 8, y = 15$,

which gives 120 for the number of balls used in space pool. It is the eighth triangular number and the fifteenth tetrahedral number.

There are only two more solutions:

$x = 20, y = 55$.
$x = 34, y = 119$.

These solutions give 1,540 and 7,140 for the number of balls. All five solutions were known in the late nineteenth century, but it was not until 1967 that a Russian mathematician first proved there are no others. The proof is difficult.

Suppose that instead of starting with the numbered balls in a triangle on the table, they are in a square formation. As before, the space version begins with the same set of balls in a tetrahedral packing. In other words, find a number that is both tetrahedral and square.

Two trivial solutions are 1 and 4. There is only one other. Can you find it?

The Supreme Ruler's plan will not work.

Consider all first-born children. One-third will be male, one-third female, one-third bisexual. Mothers who give birth to bisexuals will be sterilized.

The remaining mothers may have second children. One-third of the second-born will be male, one-third female, one-third bisexual. Again, mothers of the bisexuals will be made sterile.

The remaining mothers may have third children, and so on. This obviously generalizes to families of any size. The proportions of sexes will always be 1:1:1.

Assume that the decree lasts a thousand years and that all mothers live long enough, and are healthy enough, to keep bearing children until they have a bisexual. What will be the average number of children born to a Byronian mother during the millennial period?

The algorithm uses the binary system. Take 1 doyle from the first can, 2 from the second, 4 from the third, 8 from the fourth, 16 from the fifth, and 32 from the sixth. These numbers, 1, 2, 4, 8, . . . , are powers of 2, and every integer is the sum of a unique set of such powers, provided no two are alike.

Place the 63 doyles on the scale and write down the excess weight in milligrams as a binary number. The position of each 1 in the number, counting from the *right*, identifies a defective can. Example: The excess weight is 22 milligrams. In binary, 22 = 10110. Therefore the second, third, and fifth cans hold defective doyles.

Several months later, after a third shipment, the following message came: "Due to computer error, each can contains only two dozen doyles. Any can may be full of defective doyles, each 1 milligram overweight. Destroy defective doyles."

"The binary system won't work now," said Watts. "It requires 32 doyles from one can, but no can has that many."

Shurl said nothing. He retired to his room where he gave himself an injection of Fermataine, a drug that increases one's ability to do number theory in book margins, and scraped for a while on his musical saw. When he returned he said:

"I've done it again, Watts. One weighing suffices. A most singular solution."

What was Watts' solution?

The question is easily solved by time-reversing the event. If the container was filled after 50 hours, it was ⅐ filled after 49 hours.

That should have been easy. But consider, now, a more difficult question. Suppose Dr. Moreau III had put 7 microbes into the container instead of 1. After how many hours would the container have been ⅐ full of microbes?

7

Leo established his curious theorem by way of elementary graph theory. He placed n spots on the napkin to represent any group of n people. Each handshake can now be represented by a line connecting the two spots. Misanthropic spots will have no lines. Some spots will have only one line, and others will have many. Some pairs of spots will be multiply connected by many lines, should the same pair of people be introduced to each other over and over again. Leo's theorem is clearly equivalent to the graph theory assertion that no matter how many lines are drawn, the total number of spots with an odd number of lines will be even.

Here's one proof. Call the number of lines emanating from any point the "score" for that point. A point with an even score is an "even point," and a point with an odd score is an "odd point." Since every line joins two points, the total score for *all* the points must be even.

The total score for all the even points must also be even because any number, multiplied by an even number, gives an even product. If we now subtract this score from the even total score, we get the total score of all the odd points. Since any even number taken from an even number leaves an even number, we conclude that the total score of the odd points is even.

One final step remains. Only an even number of odd numbers can be even. Therefore the number of odd points (which we know have an even total score) must also be even. Therefore the number of persons who have shaken hands an odd number of times is even.

Ling listens carefully while Leo slowly goes through the proof. Suddenly he grins. "Your proof, my friend, has an enormous black hole in it. In fact, it's false. I've just thought of a counterexample."

"But that's impossible," snorts Leo. "The proof is airtight. There *can't* be a counterexample."

Ling proceeds to completely demolish Leo's proof. What does he do?

8

Let a be each half of the chord AC, let r be the radius of the inner circle, and let R be the radius of the outer circle. (See illustration.)

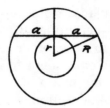

A circle's area is pi times the square of its radius. Therefore the area of the small circle is πr^2, and the area of the large circle is πR^2. The ring's area, the difference between the areas of the two circles, is $\pi R^2 - \pi r^2$, or $\pi(R^2 - r^2)$.

Since a and r are two legs of a right triangle, with R as its hypotenuse, we know (from the Pythagorean theorem) that $a^2 + r^2 = R^2$. Rearranging terms gives $a^2 = R^2 - r^2$. This allows us to substitute a^2 for $(R^2 - r^2)$ in the previous equation. The startling result is that the two unknown terms, r and R, drop out, leaving the simple formula πa^2. Since $a = 100,000$ kilometers, the area of the ring is pi times $100,000^2$, or $31,415,926,535.89+$ square kilometers.

Captain Quank worked all this out without speaking while Lieutenant Flarp mixed a pitcher of dry Martian martinis.

"I've got it!" shouted the captain. "The ring's area is—"

"Don't tell me," interrupted the lieutenant. "Let me guess. It is"—he paused to check a number he had jotted on the back of an envelope—"$31,415,926,535.8979+$ square kilometers."

"Flarp, there are times when you amaze me. You're

absolutely right. But how did you do all that algebra in your head?"

"I didn't do any algebra. All I needed was the formula pi-r-squared. I've never forgotten it because when I was a boy, and told my father I'd learned it in school, he said, 'Son, your teacher is crazy. Pie are round.' "

Second question: How did Lieutenant Flarp solve the problem so easily?

Figure 2 shows a sample chain that obviously can be enlarged to include any number of links. If one link is removed, the others are free of one another.

9

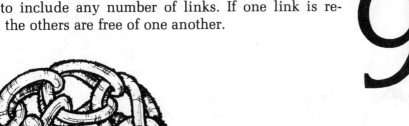

FIGURE 2

On rare occasions, a pair of toroids grow joined together like Siamese twins. Figure 3 shows two such forms, one linked and one unlinked.

Our second question: Are these two forms topologically equivalent? Think of them as surfaces made of thin rubber

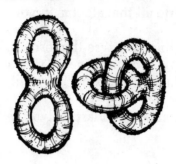

FIGURE 3

that can be stretched, shrunk, and distorted as much as you like provided the surface is never broken or torn. Is it possible, by such deformation, to change one form to the other?

10

The four stamps must have the values 1, 4, 7, and 8.

As the U.S. colony expanded, more domes were built. It soon became necessary to replace the first series of stamps with a new series of five stamps so that higher sums could be made. Tate had little difficulty proving that the best set was 1, 4, 6, 14, and 15. With one, two, or three of these stamps one can make sums of 1 through 36.

A few years later, further growth of the colony required a new series of six stamps. For this set Tate was able to show that there are just two "best" sequences, each going as high as a sum of 52 dollars:

1, 3, 7, 9, 19, 24
1, 4, 6, 14, 17, 29

Eventually it became necessary to issue a series of seven stamps. At this point the task of finding the best set of values became so difficult that Tate had to seek the help of mathematicians on earth who were experts in combinatorial number theory. They told him there was no known formula for obtaining sequences of this sort, but they wrote a computer program that made an exhaustive search and found the best sequence for seven stamps. It turned out to be unique, and to provide any sum from 1 through 70 dollars.

What is the sequence?

1. The letters in the last name of the man who formulated the famous three laws of robotics are A-S-I-M-O-V.

2. The initial letters of the planets, starting with the outermost one and proceeding toward the sun, are P, N, U, S, J, M, E, V, M.

3. Seven letters can be crossed out to leave IV PLUS IX PLUS V, which equals 18.

4. DRAW is the only verb on the list with a past tense that does not rhyme with *ought*.

5. July 4, 2000, is a date in the twentieth century. The twenty-first century begins January 1, 2001.

6. MAINE is the shortest name of a state that shares a letter in common with each of the other forty-nine states.

7. The triangle shown has a base of 8 and sides of 5 and 3. Since $5 + 3 = 8$, it is a triangle that has degenerated to a straight line. On this line the line segment labeled x has a length of 1.

8. The only English word formed by two three-letter abbreviations for a month is DECOCT.

9. The only number that meets all the specified requirements is 735.

10. Believe it or not, the long hand passes the short hand of a clock only ten times during a twelve-hour period.

FIRST ANSWERS

POSTSCRIPT

The test was, of course, intended only as a joke, but I can hardly fail to sympathize with many readers of *IASFM* who did not think it funny.

For question 6 I am indebted to the late problemist David L. Silverman. "As Maine goes," he added in a letter, "so goes the nation." Reader Rick Humburg pointed out that there is no state without at least one letter from the set *a,i,n*, and that fifteen states contain all three letters.

Question 8 is also from Silverman. Question 9, with its unique answer, was devised by the late J. A. Lindon, a British wordplay expert and writer of comic verse.

If you find the answer to question 10 hard to believe, test it with a clock or watch.

12

Our problem first appeared in *Eureka*, a publication of mathematics students at Cambridge University, in October 1966. Here is how Professor D. Mollison, of Trinity College, Cambridge, answered it.

"The three [points] are undistinguished random points. Consider each in turn as moving to its right (say) till it reaches one of the others. We see that the three distances are identically distributed random variables with sum 1; hence each has mean ⅓."

In other words, Smith and Jones can "expect" to walk a distance equal to one-third of the crater's circumference. This, of course, is an average over the long run of repeated trials.

After reaching the first supply station, Jones and Smith loaded their packs with food and began to explore the crater. First they walked in a straight line from the supply station until they reached the crater's rim again. The distance they covered was 5 kilometers. They then turned at a

90-degree angle and walked in a straight line for a distance of 12 kilometers until once more they arrived at the rim.

What is the crater's diameter?

Eighty pink arms mean 40 pinks, and half as many blue arms mean 20 blues. Thus there are 60 pinks and blues. Subtracting 60 from 81 gives 21 as the number of eyes on the greens.

Twenty-one can be factored in just two ways: 1×21, and 3×7. We were told, however, that there are more than 3 greens and less than 12, therefore there must be 7 greens, each with three eyes. All three races have a third eye located centrally just above the nose.

Now for a second problem. Turn back, reread the dialogue, and identify each professor's skin color.

13

Let T stand for truther, L for liar, and S for sometimer. There are six possible permutations:

	Left	Middle	Right
1.	T	L	S
2.	T	S	L
3.	L	T	S
4.	L	S	T
5.	S	T	L
6.	S	L	T

Go over the questions and answers, applying them to each of the six cases. Only the sixth case does not produce a contradiction. Therefore the left robot is the sometimer, the middle robot is the liar, and the right robot is the truther.

Professor Tinker congratulated the student on his solution. For a second test he asked the three ladies to leave the

14

room, then return and seat themselves again, though not necessarily in the same order. This time one of the girls was wearing an emerald necklace.

"Each robot was made on a different day," said Professor Tinker. "Therefore one of them is older than the others. All three know who she is. Your problem is to ask just *two* questions that will tell you whether the girl with the necklace is or isn't the oldest."

There was a long period of silence during which the students scribbled furiously on their notepads. Then Azik Isomorph, the brightest student in the class, raised his hand. How did Isomorph solve the problem?

---

15 The game generalizes to any odd number of spots on a zig-zag line like the one shown. The first player can always win if the number of spots (not counting the starting spots) is not 5, or the sum of 5 and any multiple of 8. The sequence of such numbers is 5, 13, 21, 29, 37, ...

On the board shown, the number of spots (excluding starting spots) is 15. This is not in the above sequence; hence the first player can always win. His first move must be to advance two spots. Thereafter he adopts one of the two following alternatives (he can always do one or the other):

1. He plays so that after his move he occupies an inside spot, and has left 1, 8, or 9 spots between the two counters, or

2. He plays so that after his move he occupies an outside spot, and has left 4, 5, or 12 spots between the two counters.

When Bagson analyzed the game he suddenly realized it was equivalent to a simpler take-away game played with a pile of counters and described in many twentieth-century puzzle books. Can you think of a way to play a game, isomorphic with the board game, that uses nothing more than a pile of pebbles and rules for removing them from the pile?

To the gnome's first pricing system, I applied the following general function: The sum of any consecutive sequence of x odd numbers starting with 1 is x^2. For example, the sum of $1 + 3 + 5 = 3^2 = 9$.

The second pricing system was more involved. To minimize the total cost of the magazines it was necessary to divide them as nearly in half as possible. Had there been an even number of magazines, the two stacks would have been equal and each magazine in one stack would have had a price duplicated by a magazine in the other. Because I told my wife that only one magazine cost five times her age, one stack must have contained one more magazine than the other.

Let x^2 be the price in dollars of the smaller stack and $(x + 1)^2$ be the price of the larger. The first stack contains x magazines, the second contains $x + 1$. The most expensive magazine in the larger stack (the only one not duplicated in price by another) costs $2x + 1$. We know that this amount is five times my wife's age. Let a be my wife's age. We can write the equation $2x + 1 = 5a$. Since the left side is clearly an odd number, then a must be odd. Rearranging terms, we can express x as equal to $(5a - 1)/2$.

The total cost of all the magazines is $x^2 + (x + 1)^2$. If for x we substitute the equivalent value of $(5a - 1)/2$, we get an expression that simplifies to:

$$\frac{25a^2 + 1}{2}.$$

This is the total cost of the magazines where a is a positive odd integer.

16

Any positive odd integer can be expressed as $2n + 1$, where n is any non-negative integer. Substituting this expression for a in the above formula gives us:

$$\frac{25(2n + 1)^2 + 1}{2}$$

which simplifies to $50(n^2 + n) + 13$. This is the total cost of the magazines in dollars where n is any positive integer. Since $n^2 + n$ is always an even number, the total cost must be a multiple of 100 dollars with 13 dollars added to it. Therefore, to raise the total to a multiple of 100, the gratuity is always $100 - 13 = 87$ dollars regardless of my wife's age or how many magazines I bought!

When I got home I discovered that the magazines were not in as good condition as I thought. One was missing pages 8, 9, 13, 27, 28, and 33. How many leaves had been torn from it?

17

Observe that the figure has five compartments. If there is a solution, the twisty line that enters a compartment from the outside, then leaves the compartment, must cross two line segments—one while entering, one while leaving. If the compartment is surrounded by four line segments, as in the case of compartments A and B, the line can enter and leave the compartment twice. However, if a compartment is surrounded by five line segments, all five can be crossed only if the line has one of its ends *inside* the compartment.

Three compartments (C, D, and E) are each surrounded by five line segments. Therefore, if the puzzle is solvable, each of these compartments must contain one end of the twisty line. But a line has only two ends. Therefore, there is no way to solve the puzzle without leaving at least one line segment uncrossed.

Couth left his daughter alone to work on the proof, then

returned later to see if she had found it. To his surprise, Tanya not only had discovered the proof—she also had found a fallacy in it! In fact, she had discovered a way to solve the original problem!

How did Tanya solve it?

Dr. Loveface thought of the following event: "ORACLE will make its next prediction by turning on its red light."

This would force the computer into a logical contradiction. If it turned on the red light for "no," the prediction would be wrong because the red light did in fact go on. If it turned on the green light for "yes," this too would be wrong because the green light went on, not the red.

While Professor Blabbage was recuperating, Dr. Loveface actually gave the event to ORACLE and requested its prediction. The computer's circuits went into a yes-no loop, producing a humming sound that grew steadily louder until suddenly the entire computer exploded, completely destroying Blabbage's life work.

There are many variations of this basic paradox which show that under certain conditions predictions of the future are impossible in principle. Can you think of an equivalent version of the computer paradox so simple that you can inflict it on a friend by speaking less than fifteen words?

19

If you tried to crack this puzzle by algebra, using exact quantities, you probably got into a muddle. There is, however, a ridiculously simple proof that the amount of blood in the vodka must exactly equal the amount of vodka in the blood.

We are told that at the finish there was, as before, one quart of liquid in the large bottle, one pint in the small bottle. Consider the large bottle. It is missing an x amount of vodka. Since it remains a quart, this missing amount must have been replaced by an x amount of blood! Of course the same reasoning applies to the small bottle. If it is missing an x amount of blood, and remains a pint, the missing blood must be replaced by an x amount of vodka. In fact, it doesn't matter in the least how many times varying amounts of liquid are transferred back and forth so long as at the finish there is a quart in one bottle and a pint in the other. Even the bottle sizes are irrelevant. The vodka in the blood must equal the blood in the vodka!

Can you invent a simple card trick based on the same curious principle?

20

Philbert reasoned as follows:

"Suppose my erasure day is Saturday. No one will tell me Friday morning that Saturday is the day, therefore on Friday afternoon I will *know for certain* that the day is Saturday. But the judge told me I would *not* know the day until the morning of the day itself. Therefore I *can't* be erased on Saturday without making the judge a liar.

"Consider Friday. It too is ruled out. Since Saturday cannot possibly be the day I stick my head in the oblivion box, if I'm not told the day by Thursday noon, I will know that the day is Friday. Why? Because only Friday and Saturday remain. It can't be Saturday, hence it must be Friday. But if I know on Thursday that it is Friday, the judge again will have uttered a falsehood.

"So Friday and Saturday are out. Consider Thursday. It too is eliminated by the same reasoning! After twelve o'clock on Wednesday, if I've not been told the day, I will know it is Thursday because it can't be Friday or Saturday. The same reasoning applies to Wednesday, Tuesday, and Monday. No matter what day is picked, I'll know the date by the afternoon of the previous day. In each case it will make the judge a liar and allow me a new trial."

Philbert's reasoning seems impeccable, yet there is a fatal flaw in his logic. It is not so easy to pinpoint exactly where the flaw lies, but it *is* easy to prove that Philbert's reasoning can't be correct. How?

---

1. H. G. Wells, "The Country of the Blind."

2. Robert Heinlein, "And He Built a Crooked House."

3. Lester del Rey, "Evensong."

4. Arthur C. Clarke, "The Nine Billion Names of God."

5. L. Sprague de Camp, "The Command."

6. Isaac Asimov, "Nightfall."

21

Members of the Oulipo are fond of anagrams—words, phrases, or sentences with letters rearranged to make other words, phrases, or sentences. Not many last names of writers will form anagrams of common English words, but some of the above will. *Wells*, for instance, has one anagram: *Swell*. *Clarke* has at least three: *Calker*, *Lacker*, and *Rackle*. *Asimov*, *Heinlein*, and *del Rey* seem hopeless, though *O del Rey!* anagrams to *Yodeler*.

Can you find a single-word anagram for *de Camp*?

22

The letter T is at positions 2, 3, 5, 7, 11, 13, 17, 19, 23, 29, 31. These are the first eleven prime numbers.

"The primes certainly prove the virus is artificial," said Witson. "But perhaps there's another message. The number thirty-six suggests a six-by-six square matrix. Let's try scanning it. We can color each cell with one of four colors, a color for each letter. Maybe a geometrical picture will turn up."

To their delight, a picture did indeed turn up. What was it?

23

In her mind, Tanya just turned the small triangle upside down:

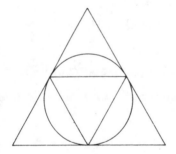

It was obvious at once that the large triangle consisted of four of the small ones, therefore its area was four times that of the small triangle.

Now for a slightly more confusing problem. Suppose the figure had been a circle inscribed in an equilateral triangle, and the triangle in turn inscribed in a larger circle:

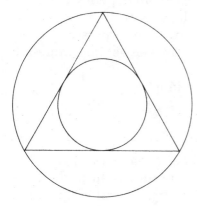

What is the ratio between the areas of the two circles?

Here is how Ada Loveface proved that if tachyons exist they can't be used for sending signals with speeds faster than light:

Suppose that A, at Barkback College, is in communication by a tachyonic antitelephone with B, who lives on a planet at the other side of our galaxy. The tachyon speeds and the distance are such that if A sends a signal to B, and B instantly replies, then B's signal will arrive one hour before A sends his signal. A could then get an answer to a question an hour before he asked it!

Dr. Loveface sharpened the contradiction as follows. Suppose A and B agree that A will ask his question at noon if and only if B's immediate reply does not reach him by 11 A.M. on the same day. We are forced to conclude that an exchange of messages will take place if and only if it does not take place—a flat logical contradiction.

A few days after Professor Cracker had abandoned his project, Ada approached him and said: "Perhaps I was too hasty the other day in saying your antitelephone couldn't work. I've been reading some old science fiction classics about time travel, and they have suggested a possible way a modulated tachyon beam *could* send a signal that wouldn't lead to an absurdity."

What does Dr. Loveface have in mind?

1. Rocket ships are in "free fall" as soon as they leave the earth. From the time the motors are turned off to the time they are used again for altering course or braking, there is zero g inside a rocket ship.

2. Cigars won't stay lit in zero gravity unless you constantly wave them about. Gasses produced by the burning of tobacco must be carried upward by the buoyancy of air, in turn caused by gravity pulling air down.

3. Birds can't fly on the moon because there is no air against which their wings can push or support them when gliding.

4. No air, no breezes, no rippling flags on the moon.

5. Although in daytime the lunar sky is indeed dark, there is so much reflected light from the moon's surface that stars are not visible to unaided eyes. They *can* be seen through binoculars.

6. Even at night, stars on the moon never twinkle. Twinkling on earth is caused by movements of the atmosphere.

7. For stars to be visible inside the arms of a crescent earth they would have to be between earth and the moon.

8. The moon does rotate once during each revolution around the earth, but since it always keeps its same face toward the earth, the earth does not rise and set. From any given location on the earth side of the moon, the earth remains fixed in the sky.

9. Without air a boomerang can no more operate on the moon than a bird can keep itself aloft.

10. Twitchell couldn't have heard the boomerang strike the boulder because sound requires an atmosphere to transmit its waves to a human ear.

11. Before the first moon landing it was widely thought that objects would be invisible in moon shadows. Actually, so much light is reflected from the irregular lunar surface that this is not the case.

12. Although the sun does rise and set on the moon, it takes it about twenty-eight days to return to a former position. It could not have set as rapidly as the narrative indicates.

13. The terminator moves at about 10 miles per hour. This is slow enough for a person to keep pace with its movement.

14. Meteors leave glowing trails only when they are burned up by friction of the earth's atmosphere. On the atmosphere-less moon, meteors would not produce such trails.

15. As in mistake 12, the sun could not have risen until some two weeks after it set.

POSTSCRIPT

One reader questioned mistake 10. Sound travels through any medium except a vacuum, so if Twitchell's suit did not insulate him against sound, he may have heard the boomerang fall. I doubt it. At the most he might have felt a slight vibration with his feet. Other readers pointed out that if there are convection currents in a spaceship, cigars might stay lit in zero gravity. I wonder. Cigars go out easily even in normal gravity.

Edgar D. Twitchell is a play on Edgar D. Mitchell, the astronaut. Since his walk on the lunar surface, he has devoted all his time and energies to encouraging research on the luniest aspects of contemporary psychic research, such as metal bending by the mind, and the paranormal perception of plants.

Some of the bloopers in the puzzle have been made by the most eminent of SF writers. The first one was made by Jules Verne in his novel *From the Earth to the Moon*. A second blooper in the same novel concerns a dog on the ship. When the dog dies, its body is flung out a window where it remains alongside the ship during its journey. Of course anything flung out the window of a spaceship would continue to move away from the ship.

In one of her SF stories, Judith Merrill makes a blunder similar to the mistake about the birds and boomerangs. She has helicopters flying about on the moon. See the entry on "Scientific Errors" in *The Science Fiction Encyclopedia*, edited by Peter Nicholls.

26

A doubly-weird sequence must begin as shown below. Remember that not only must each horizontal row have numbers in increasing order, but the "leading edge"—the first slanting row formed by the initial numbers—must also be an increasing sequence. In this case the leading edge begins 1, 2, 4, ...

```
1 3 9  20  38  64  100 . . .
 2 6 11  18  26  36 . . .
  4 5  7   8  10 . . .
```

As you proceed to expand the sequence, every number is forced. All goes well until you reach the tenth number of the top row, which is 284. Unfortunately the nineteenth number of the second row also is 284.

How about triply-weird numbers? In other words, is there a sequence of increasing natural numbers, with three rows of increasing differences, and an increasing leading edge of first numbers for each row, that contains all the counting numbers without duplication?

The bet strongly favors the Devil. No matter what triplet the man picks, the Devil can pick a triplet that is likely to win with odds that range from 2 to 1 to 7 to 1!

It is true that for any given set of three plays the probability one triplet will show is the same as any other. But the probability of which of two triplets will show first is an altogether different matter. Here's how it breaks down. Let R stand for red, B for black. The left column lists the eight triplets the man can choose, the middle column gives the Devil's best response to each, and the third column gives the probability that the Devil will win.

Man's choice	Devil's choice	Probability
R R R	B R R	$7/8$
R R B	B R R	$3/4$
R B R	R R B	$2/3$
R B B	R R B	$2/3$
B R R	B B R	$2/3$
B R B	B B R	$2/3$
B B R	R B B	$3/4$
B B B	R B B	$7/8$

The chart answers our second question. Although the man cannot pick a triplet that gives him an advantage, he has the best chance of winning (one out of three games) if he picks *RBR*, *RBB*, *BRR*, or *BRB*. His worst choices are *RRR* and *BBB*. They allow the Devil to win seven out of eight times in the long run!

On another evening in Vegas the Devil's victim was a Harvard economics student, on vacation with his girl friend.

"The probability of the ball dropping into any specified slot," said Satan, "obviously is $\frac{1}{38}$ because the numbers go from 1 through 36 and the wheel has a zero and a double zero. Suppose we ask the croupier to give the wheel a spin just for us, and to use two balls instead of one. What's the probability that both balls will wind up in the same slot?"

"Let's see," said the student. "If I remember my probability theory correctly it's $\frac{1}{38}$ times $\frac{1}{38}$, or $\frac{1}{1,444}$."

"Right," said the Devil, smiling. "So how about you betting 100 ozmufs against my single ozmuf that the balls will drop into different slots."

Whom does the bet favor this time?

28

You are riding on ship D. Try pushing counters if you don't believe it.

After you understand this puzzle, here is a closely related one, though a trifle trickier. At noon the long and short hands of your watch coincide at twelve. At exactly what time, expressed to the fraction of a second, will the minute hand next coincide with the hour hand?

29

The palindromic sequences are:

NO LEMONS, NO MELON.
PULL UP IF I PULL UP.
NO, IT IS OPEN ON ONE POSITION.

Now focus your mind vigorously on this paragraph and on all its words. What's so unusual about it? Don't just zip through it quickly. Go through it slowly. Tax your brain as much as you can.

1. If you add the digits of any number, then add the digits of the sum, and continue in this way until one digit remains, that digit is called the "digital root" of the original number. The year 2556 has a digital root of 9, and of course scrambling digits cannot change a digital root.

A curious property of any number with digital root 9 is that all its multiples are digital root 9. Adding 100 (digital root 1) to the multiple produces a number with digital root 1. The digital root of any number is the same as the remainder when that number is divided by 9 (with the exception of numbers that are digital root 9, in which case there is no remainder).

When your final result is divided by 9 there will be a remainder of 1; and since 3 goes evenly into 9, the remainder will also be 1 when the number is divided by 3. Therefore the quotient will have a fraction of ⅓. In your calculator's readout this will appear as the repeating decimal fraction .33333 . . .

2. The only way to insert the nine positive digits into the pattern to obtain the indicated sums is:

```
  963
  852
  741
-----
 2556
```

3. The only way to insert plus or minus signs into the sequence 1 2 3 4 5 6 7 8 9 to obtain a sum of 556 is:

$$1 - 2 + 3 + 4 + 567 - 8 - 9 = 556.$$

4. The only way to insert three plus or minus signs into the same sequence to get a sum of 56 is:

$$123 - 45 + 67 - 89 = 56.$$

5. My best efforts to express 56 with three fours or fewer are:

$$(4! + .4) \ 4 = 56; \quad \text{and} \quad \sqrt{4} \ (4! + 4) = 56.$$

6. If numbers are placed at the left of each row and above each column as shown below, you will see that every cell contains the sum of the pair of numbers that mark its row and column. Every selection of a cell number eliminates just one pair of the "generator" numbers. Since the ten generators add to 56, it follows that the five circled numbers must also add to 56.

	2	4	5	1	3
8	10	12	13	9	11
7	9	11	12	8	10
11	13	15	16	12	14
9	11	13	14	10	12
6	8	10	11	7	9

POSTSCRIPT

Here's an easy proof of the uniqueness of the solution to the second problem. Only the three highest digits (9, 8, 7) will give a sum as high as 24. Only the next three highest digits (6, 5, 4) will add to 15. This supplies the 5 below and the 1 to carry to raise 24 to 25. Numbers 3, 2, 1 remain for the last column. Only 9, 6, 3 in the top row add to 18, and only 8, 5, 2 in the middle row add to 15, leaving 7, 4, 1 for the bottom row.

At first I marveled at Rendrag's ingenuity in finding so many problems involving 2556 that had unique answers. But after thinking more about the matter I concluded it was not so difficult after all. For example, suppose the year had been 2223. The problems can be modified as follows:

1. The calculator trick is unchanged. Simply start by scrambling 2223 instead of 2556.

2. The same as before except the sums on the right (top down) are 15, 12, and 9. The total of the three three-digit numbers is 2223, and the digits to be used are 0 through 8. The unique solution is:

$$
\begin{array}{r}
852 \\
741 \\
630 \\
\hline
2223
\end{array}
$$

3. Insert plus or minus signs into the descending sequence 9 8 7 6 5 4 3 2 1 to make a sum of 223. The only answer is:

$98 + 76 + 5 + 43 + 2 - 1 = 223.$

4. Insert just five signs (plus or minus) in the same sequence to make a sum of 23. Again there is only one answer:

$9 + 8 + 7 - 65 + 43 + 21 = 23.$

5. Express 23 with no more than three fours, and without using the decimal point or radical sign that were permitted before. The only answer I know is:

$4! - (4 \div 4) = 23.$

With four fours and radical signs it can be done this way:

$$4! - \frac{\sqrt{4} + \sqrt{4}}{4}$$

6. Here is a square that forces 23:

```
  1  3  0  2
4 5  7  4  6
2 3  5  2  4
6 7  9  6  8
5 6  8  5  7
```

For more details on how to construct these mystifying squares (they can be based on multiplication as well as addition), see chapter 2 of my first collection of columns: *The Scientific American Book of Mathematical Puzzles and Diversions*.

31

Bob said to himself: "There are just two possibilities. The opaque box is either empty or it contains the check. Suppose it's empty. If I take only the opaque box I get nothing. But if I take both boxes I get at least a hundred dollars. Suppose the opaque box is not empty. If I take only it, I get the million-dollar check. But if I take both boxes I get the check *plus* a hundred dollars. Either way I'm sure to come out a hundred dollars ahead by taking both boxes!"

Each argument seems impeccable. According to experts on decision theory, which one is right?

32

The clerk simply requested over the laser intercom that the occupants of each room move next door to the room with the next higher number. This left room 1 vacant for the creature from Andromeda.

A few days later, when the inn was still filled to capacity, ten humanoid couples on a parallel-universe tour arrived at the inn. Each couple had a reservation for a separate room.

The inn had, of course, no difficulty taking care of them. The clerk merely moved everybody to a room with a number ten higher than the number of the room they were in. This left rooms 1 through 10 vacant for the ten couples.

In a similar manner the inn found rooms for thousands of guests with reservations who arrived during the next twelve days, even though all rooms were occupied when each new guest arrived.

On the thirteenth day, however, an unusual event occurred for the first time in the inn's history. A convention had been scheduled for science fiction buffs from *all* the parallel worlds associated with the Black Tube, and an infinite number of fans showed up. Every single one had a reservation.

At first the clerk was puzzled over how to accommodate such a large set, but one of the fans, a youngster who somewhat resembled a purple ostrich, had been studying Cantorian set theory in school. "It's ridiculously simple," he said to the kangaroo. "All you have to do is . . ."

What suggestion did he make?

33

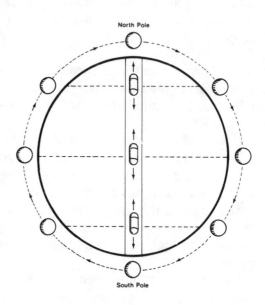

1. The car's velocity steadily increases from zero at the start to maximum at the earth's center, and steadily decreases thereafter to zero at the other end.

2. The car's acceleration is maximum at the start (32 feet per second per second). It decreases as it approaches the earth's center, where it becomes zero. After that it accelerates negatively until it reaches the other end.

3. Halfway down the tube, in a stationary car, you would weigh much less than on the earth's surface because of the gravitational pull of the earth above you.

4. You would be in free-fall throughout the entire trip, and therefore always in a state of zero gravity.

5. The car reaches a top speed at the earth's center of about 17,700 mph, or almost 5 miles per second.

6. On the moon a car falling through the moon's center

would complete the trip in about 53 minutes; on Mars, in about 49 minutes.

7. The story is "When the Earth Screamed," by Sir Arthur Conan Doyle. It tells how Professor George Edward Challenger, the hero of Doyle's novel *The Lost World*, penetrates the earth's "skin," causing the earth to howl with pain.

POSTSCRIPT

A tube that goes straight through the earth's center has been the basis of many SF stories and novels. Plutarch seems to have been the first to ask what would happen to a body that fell through such a tube, and Galileo apparently was the first to answer correctly. In eighteenth-century France, Voltaire and astronomer Pierre Louis Moreau de Maupertuis argued over the question.

The earliest instance I know of the tube's use in a SF novel is *Through the Earth* (1898) by Clement Fézandié, a New York City public school teacher. His short stories about "Dr. Hackensaw's Secrets" appeared regularly in Hugo Gernsback's *Science and Invention* before Gernsback started *Amazing Stories* in 1926, and I have often wondered why these more than forty amusing tales have never been gathered in a book. *Through the Earth* was first serialized in *St. Nicholas* magazine, volume 25, in four installments from January through April 1898.

In Fézandié's novel, the tube is drilled simultaneously from the United States and Australia, using electricity supplied by tidal energy. A cooling system in the tube counteracts the earth's intense interior heat, and the tube is lined with a new heat-resistant metal called carbonite. A vacuum is maintained inside the tube to eliminate air resistance. Electronic repulsion prevents friction between the sealed car and the tube's sides. William Swindon, sixteen years old, volunteers as the first passenger, but you'll have to look up the serialization or locate a copy of the rare book to learn what happens on the trip.

FIRST ANSWERS

In 1929 Appleton published *The Earth-Tube* by Gawain Edwards, a pseudonym of rocket expert Gawain Edward Pendray, about a war between the United States and Asia. The Asiatics, after boring a hole through the earth and lining it with a metal called "undulal," pour men and undulal tanks into the tube to conquer the Americas after they emerge near Buenos Aires. The plot is foiled by the United States' discovery of a way to destroy undulal.

Shorter tubes that go straight from one city to another have also been used in SF for transportation. Neglecting friction and air resistance, no fuel is needed for a train because gravity draws it to the middle of the tunnel, then momentum carries it the rest of the distance. This was the basis of Alexander A. Rodnykh's novel, *Subterranean Self-propelled Railroad between St. Petersburg and Moscow*, published around 1900, and a 1913 German novel, *Der Tunnel* (English Translation: *The Tunnel*, 1915) by Bernhard Kellermann, concerning a similar tube from New Jersey to France. The idea of using gravity to start and brake a car is actually employed now in many subway systems by putting curves at the beginning and end of stops, and we are all familiar with the principle's use in bowling alleys for returning balls to the bowler.

The German Professor in Lewis Carroll's *Sylvie and Bruno Concluded* (1893) explains to Lady Muriel how the straight tunnel permits a gravity train. L. Frank Baum uses a gravity tube for transportation in *Tik-Tok of Oz*.

If we assume a homogenous earth, ignore air resistance, friction, Coriolis forces, and so on, it is not hard to calculate that a car falling straight through the earth's center would make the trip in a trifle more than 42 minutes. Surprisingly, this time is independent of the tube's length. No matter how short, in a tunnel that goes straight from one spot on the earth's surface to another, the time for a trip is about 42 minutes, or 84 minutes for a round trip.

It is no coincidence that the falling body's maximum

speed is precisely the speed (it was calculated by Newton) at which a satellite must be fired horizontally to put it in a circular orbit just above the earth. Under ideal conditions (no atmosphere, spherical earth, and so on) the satellite would complete one orbit in about 84 minutes.

Imagine the earth's axis perpendicular to the plane of the ecliptic, and the satellite circling the earth from pole to pole on a plane that intersects the sun. Further imagine that the sun casts a shadow of the satellite on the earth's axis. The shadow would oscillate back and forth from pole to pole in exact conformity with the oscillation of a gravity train—an internal satellite!—inside a tube from pole to pole. This is a way of saying that the train would oscillate with simple harmonic motion. Indeed, a gravity train on a straight track of any length through the earth would oscillate with harmonic motion.

It also is no coincidence that 84 minutes is the period of the so-called Schuler pendulum, an imaginary giant pendulum as long as the earth's radius and swinging just above the earth's surface.

The following selected list of references, chronologically ordered, will probably tell you more than you care to know about gravity tubes through the earth:

Flammarion, Camille. "A Hole Through the Earth." *Strand Magazine*, vol. 38, September 1909, pp. 349–55.

Lindgren, Harry. "Subterranean Travel." *Australian Mathematics Teacher*, vol. 9, 1953, pp. 34–9.

Cooper, Paul W. "Through the Earth in Forty Minutes." *American Journal of Physics*, vol. 34, January 1966, pp. 68–80.

Stretton, William C. "Straight-line Tunnels Through the Earth." *Mathematics Teacher*, January 1967, pp. 12–13.

Cundy, H. Martyn. "Quicker Round the Bend!" *Mathematical Gazette*, December 1968, pp. 376–80.

"Fast Tunnels Through the Earth," problem 5614, answered in

American Mathematical Monthly, vol. 76, June-July 1969, pp. 708–9.

Bullen, K. E. "The Earth and Mathematics." *Mathematical Gazette*, vol. 54, December 1970, pp. 352–53.

Lindgren, Harry. "Quicker Round the Bend!" *Mathematical Gazette*, June 1971, pp. 31–21.

34

1. 198 seconds.

2. I never saw a purple cow.
     I never hope to see one.
   But I can tell you anyhow,
     I'd rather see than be one.

   The author is Gelett Burgess.

3. A stitch in time saves nine.
   He who hesitates is lost.
   A rolling stone gathers no moss.
   A bird in the hand is worth two in the bush.
   Look before you leap.
   There's no place like home.

   Someone once suggested that Baum may have invented the word *Oz* by taking the abbreviation of his home state, N.Y., then shifting each letter one step forward in the alphabet. After Baum died, the Oz books were continued by Ruth Plumly Thompson, who lived in Pennsylvania. What remarkable coincidence links *her* home state with Oz?

35

"We are arranged," said Zero, "so our names are in alphabetical order. Of course that's a joke, and I wouldn't blame your readers for being annoyed. So here's a serious problem. It's never been published."

All the digits left the center of the room except two, who stood solemnly side by side. "That," said Zero, "is the smallest number with the following interesting property. Every digit from 1 through 9 is in a divisor of the number. Divisors include 1 and the number itself. Guess the number."

36

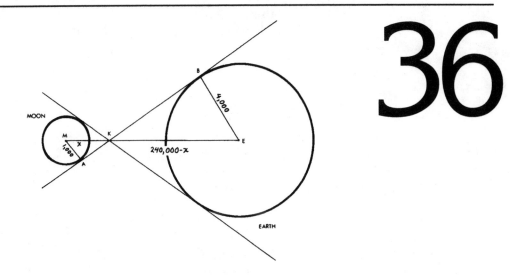

The illustration, obviously not drawn to scale, shows the earth and moon, and the line along which the Bagel is traveling. The spot on this line where the disks of the earth and moon appear identical is clearly *K*, the intersection of the two internal tangents.

Right triangles *MAK* and *EBK* are similar (having two

angles in common), therefore their corresponding sides are in the same ratio. This permits the simple linear equation:

$$x/(240{,}000 - x) = {}^{1{,}000}\!/_{4{,}000} = \tfrac{1}{4},$$

which gives x a value of 48,000 miles. At a distance of 48,000 miles from the moon's center, the earth and moon will appear identical in size.

Colonel Couth looked over Tanya's diagram and nodded approval. "Excellent. Now you can use the same drawing to work on a tougher problem. How would you go about constructing an orbit for the *Bagel*, around the moon and on a plane passing through the centers of the earth and moon, such that *anywhere along that orbit* the moon and earth will always look the same size?"

It took longer, but Tanya solved this also. Can you?

SECOND ANSWERS

Dr. Ziege could have started from a spot so near the south pole that when she made the drive east it would have taken her around the pole *twice*. Of course this generalizes to eastern trips that go *n* times around the pole, where *n* is any positive integer, so the problem is solved by an infinity of spots, on an infinity of circles.

POSTSCRIPT

A more familiar story line for this puzzle tells of an explorer who looks due south and sees a bear 100 yards away. The bear walks 100 yards due east while the explorer stands still. The explorer then points his gun due south, fires, and kills the bear. What color is the bear?

The answer is white. The bear is a polar bear, and the location is the North Pole, but as we have seen, the man may also be near the South Pole. Benjamin Schwartz, writing on "What Color Was the Bear?" (*Mathematics Magazine*, vol. 34, September–October 1960, pp. 1–4), found still other infinite sets of answers that arise from ambiguity in the problem's wording. For these solutions, and some amusing correspondence on the problem, see my *Mathematical Carnival*, chapter 17.

Does 124C41+ look familiar? Historians of SF will recognize it as part of *Ralph 124C41 +*, the title of one of the worst SF novels ever published, although astonishingly accurate in its scientific predictions. The author was none other than Hugo Gernsback, the father of SF. It was Gernsback who, in New York City, began publishing *Amazing Stories* in 1926, the world's first magazine devoted exclusively to SF. The "Hugos" awarded annually to SF writers honor Gernsback's first name. His novel ends with superman Ralph pointing out to his true love that his surname can be interpreted to mean "one to foresee for one."

2

Dr. Xenophon wears both pairs of gloves. Sides 1A and 2B may become contaminated, while 1B and 2A remain sterile.

Dr. Ypsilanti wears the second pair, with sterile sides 2A against his hands.

Dr. Zeno turns the first pair inside out and then puts them on with sterile sides 1B against his hands. Then he puts on the second pair, with sides 2A over sides 1A and sides 2B outermost.

Because only sides 2B touch Ms. Hooker in all three operations, she runs no risk of catching the Barsoomian flu from any of the surgeons.

POSTSCRIPT

As far as I know, this was the first publication of an amusing problem that had been circulating privately among mathematicians for many years. The original problem, of unknown origin, concerned three male mathematicians who were attending a conference on combinatorics. Together they took an evening off to visit a local prostitute. They were uncertain of one another's health, as well as the health of the lady. Between them they had only two prophylactics. How could the two devices be used for their protection?

The problem was soon extended, also by an unknown person, to include protection for the lady. Eventually it was generalized to n men and m ladies who paired off in all possible male-female ways. What is the minimum number of devices that provide safety for everyone? Whether this general problem has been solved or not, I do not know.

In 1977 a partial solution was found by Richard Lipton, a computer scientist at Yale University, at the same time, independently, by two Hungarian mathematicians, A. Hajnal and L. Lovász. Hajnal and Lovász titled their paper, "An Algorithm to Prevent the Propagation of Certain Diseases at Minimal Cost." The abstract of their paper follows:

"Given n rabbits $R_1, \ldots, R_n$, each having a different disease, and m plates $P_1, \ldots, P_m$, smeared with different radioactive isotopes, we want to carry out nm experiments of placing each rabbit on each plate. Protective membranes have to prevent the rabbits from infecting each other or a plate and the isotopes from getting on another plate or on an animal. In each experiment, we may put an arbitrary number of membranes on the plate under the rabbit. The same membrane may be used more than once, but if an infected surface touches another surface, an animal, or a plate, then the infection carries over. The problem is to design the order of the experiments and the use of the membranes so as to minimize the total number of membranes used. Restricting ourselves to the case $n = m = 6k$, we present an algorithm using $7k + 1$ membranes and prove that every algorithm needs at least $7k$ membranes."

Aside from 1 and 4, the only number that is both square and tetrahedral is $140^2 = 19,600$. It is the forty-eighth tetrahedral number.

3

POSTSCRIPT
If n unit spheres can be arranged on the plane to form a regular polygon, n is called a polygonal number. If they can be arranged in space to form a regular polyhedron, n is called a polyhedral number. In general, if n unit spheres form any specified plane or solid figure with pleasing symmetry, n is known as a figurate number. The ancient Greeks were fascinated by figurate numbers, and in later centuries number theorists worked on thousands of unusual Diophantine problems involving such numbers.

Our space pool problems deal with only three kinds of figurate numbers: triangular, square, and tetrahedral. Let us add a fourth, which we will call "square pyramidal" because the spheres can be piled to form a square-based pyramid like the Great Pyramid of Egypt. There are six ways to select pairs of these four number types. In each case we ask how many numbers are figurate in both senses. When the square and tetrahedral problem was solved in 1967 it almost completed the solutions for all six pairs. The numbers 10, 120, 1540 and 7140 had long been known, but not until 1967 was a proof published that there are no others. (The proof, found by a Russian named Avanesou, appeared in *Acta Arithmetica*, vol. 12, 1967, pp. 409–19, written in Russian.)

Here is what is known about the six pairs, in each case omitting 1 because it is trivially a figurate number of any type:

1. Triangular and square: an infinite sequence starting 36, 1225, 41616, ...

2. Triangular and tetrahedral: 10, 120, 1540, 7140.

3. Square and tetrahedral: 4, 19600.

4. Square and square pyramidal: 4900.

5. Tetrahedral and square pyramidal: none. As far as I know this was first proved by Raphael Finkelstein in his paper "On a Diophantine Equation with No Nontrivial Integral Solution," *American Mathematical Monthly*, vol. 73, May 1966, pp. 471–77.

6. Triangular and square pyramidal. Richard Guy has called my attention to page 255 of *Diophantine Equations*, by Louis Joel Mordell, where it is shown that the type of equation for this problem has a finite number of solutions. Only three are known: 55, 91, and 208335.

The Diophantine equation is:

$$\frac{m(m + 1)}{2} = \frac{r(r + 1)(2r + 1)}{6},$$

where the left side is the formula for triangular numbers, and the right side is the formula for square pyramidal numbers. The equation simplifies to:

$$3m^2 + 3m = 2r^3 + 3r^2 + r.$$

The known solutions, aside from the trivial $m = 1, r = 1$, are $m = 10, 13,$ and 645, and r (respectively) $= 5, 6,$ and 85. Guy conjectures that there are no others.

There is a vast literature on Diophantine analysis, of which the latest and best reference is the book by Mordell cited above. For an introduction to elementary recreational Diophantine problems see my *Scientific American* column for July 1970. The most famous of all unsolved Diophantine questions is the "last theorem" of Fermat, which conjectures that the equation $a^n + b^n = c^n$ has no integral solution for n greater than 2.

Let n be the number of mothers during the thousand-year period.

$n \times 1 = n$ children will be first-born,
$n \times \frac{2}{3} = 2n/3$ children will be second-born,
$n \times \frac{2}{3} \times \frac{2}{3} = 4n/9$ children will be third-born,
$n \times \frac{2}{3} \times \frac{2}{3} \times \frac{2}{3} = 8n/27$ children will be fourth-born, and so on.

The total number of children will be:

$n + 2n/3 + 4n/9 + 8n/27 + \cdots.$

The limit of this sum is $3n$. There are n mothers, therefore the average number of children per mother is $3n/n = 3$.

There is no need, however, to get involved with an infinite sequence that converges. Can you think of a simple solution that avoids algebra altogether?

5

From the six cans remove 11, 17, 20, 22, 23, and 24 doyles. Every subset of these six numbers has a different sum. This makes it easy to identify all defective cans in one weighing. For instance, suppose the scale registers an overweight of 53 milligrams. The only way to obtain 53, as a sum of distinct numbers in the set of six, is 11 + 20 + 22. This shows that cans one, three, and four hold the overweight doyles.

POSTSCRIPT

Shurl and Watts are, of course, burlesques of Conan Doyle's Sherlock Holmes and Dr. Watson. Herb Monroe, an *IASFM* reader, suggested some ways to improve the solutions. I quote from his letter, which appeared in the September-October 1978 issue of the magazine: "In the first solution, Shurl could have removed 0, 1, 2, 3, 4, and 5 doyles from the cans and still found the can of defective doyles. The biggest problem is that the two men are making more work for themselves or else they are throwing away valuable doyles. In the third solution, the two must discard an entire can of doyles (good or bad), or sort through a minimum of 93 doyles to keep from throwing out perfectly good doyles. . . . I think that in the long run, the two of them would be better off to remove one doyle from each can and weigh it separately—less work that way. . . ."

6

If you divided 49 by 7 to get an answer of 7 hours, you score zero. One microbe becomes 7 after the first hour. From there on, the sequence is the same as before. Therefore the container is $\frac{1}{7}$ full after 49 − 1 = 48 hours.

Now for a third question. Suppose Dr. Moreau III put just 2 microbes in the empty container. After how many hours will it be *at least* $\frac{1}{7}$ full?

Ling grasps his right hand with his left, bows, and says, "Ah so."

Although shaking hands with oneself in the Chinese style is indeed a counterexample to the handshake theorem, it is not a counterexample to the corresponding graph theorem. Can you explain why?

The lieutenant reasoned: "I trust the captain when he says the ring's area is a constant, given the length of the chord. If that's true, it makes no difference how large or small the inner circle is. Let's reduce it to the minimum—a point of zero radius. The chord is then the diameter of the outer circle and the 'ring' is the circle itself. Therefore its area is pi times the square of its radius.

"So," continued the lieutenant, "all I had to do was multiply pi by 10,000,000,000. That was easy because it just meant shifting the decimal point of pi ten places to the right."

"Beautiful!" exclaimed the captain. "But how 'in the name of Asimov can you remember pi to fourteen decimal places?"

Lieutenant Flarp handed the captain a crimson martini, then elevated his own glass and said, "How I want a drink! Alcoholic, of course, after the heavy chapters involving quantum mechanics."

Third question: How did Flarp remember pi to fourteen decimal places?

9

The two Siamese forms are topologically identical. To prove this, imagine the linked form deformed to a sphere with two "handles" as shown below.

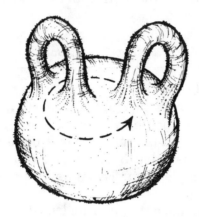

Now imagine the base of one handle being moved over the surface, by shrinking the surface in front and stretching it in back, along the path shown by the dotted line. This links the two handles. The structure is now easily altered to correspond with the linked form of the Siamese toroids.

POSTSCRIPT

James Timourian sent me this problem. He had known about the Siamese toroids puzzle for some fifteen years but could not trace its origin. For another problem about two apparently linked toroids, even more counter-intuitive, see my *Scientific American* column for December 1979 and the answer in the following issue.

David Rorvik was my inspiration for David Klonefake. His book on cloning, *In His Image*, was the most outrageous volume of pseudobiology to be published in the 1970s.

10

The only set of seven values for seven postage stamps, from which one, two, or three stamps can be selected to make any desired sum from 1 through 70, is:

    1, 4, 5, 15, 18, 27, 34.

POSTSCRIPT

A good recent reference on the generalized postage stamp problem is "A Postage Stamp Problem," by Ronald Alter and Jeffrey Barnett, in *American Mathematical Monthly*, March 1980, pages 206–210. The paper's bibliography lists forty-seven earlier references.

By "generalized" I mean this: Find all sequences of n values, for no more than m stamps to be used at one time, to obtain a sum of all consecutive values from 1 through k, where k is as large as possible. The general problem remains unsolved.

In the unlikely case you wondered what Philo Tate's joke was about snew, the victim is supposed to ask "What's snew?" To this you reply, "Not much. What's new with you?"

12

A familiar theorem of plane geometry says that if a right angle is drawn inside a circle, with its vertex on the circumference, its sides must intersect the circle at the end points of a diameter. Therefore the distances of 5 and 12 kilometers are the sides of a right triangle. Applying the Pythagorean theorem, we find that $5^2 + 12^2 = 13^2$, therefore the crater's diameter is 13 kilometers.

"By the way," said Ms. Jones. "We both know that CARTER and CRATER are anagrams. It just occurred to me that there are two English words, one hyphenated, that are other anagrams of CARTER."

What words did Ms. Jones have in mind?

13

Pink's first remark prompted a reply by the man with a blue hand, therefore Professor Pink is not a blue. Nor can he be a pink because then his name and skin color would match. Therefore Pink is a green.

The man with the blue hand cannot be Blue or Pink, therefore he is Professor Green.

This leaves pink for the skin of Professor Blue. Pink is green, Green is blue, and Blue is pink.

Now see if you can:

1. Find an anagram for GREEN, and

2. Change PINK to BLUE by altering one letter at a time so that after each alteration you get a familiar four-letter word. You must do it in no more than ten alterations.

14

Ask the robot on the left, "Is it true that the middle robot is the liar or the robot on the right is the truther?" The chart below shows the possible answers for each of the six permutations:

	Left	Middle	Right	Yes	No
1.	T	L	S	X	
2.	T	S	L		X
3.	L	T	S	X	
4.	L	S	T		X
5.	S	T	L	X	X
6.	S	L	T	X	X

As you can see from the chart, if the robot says yes, the middle robot must be a truther or a liar. If the robot says no, the robot on the right must be a truther or a liar.

If the answer is yes, say to the middle robot: "If I were to ask you if the lady with the necklace is the oldest, would you say yes?" Assume the girl with the necklace is indeed the oldest. The truther will say yes and so will the liar! (The liar would have answered no to the first part of the question, so she must lie and say yes to Isomorph's entire question.) By similar reasoning, if the girl with the necklace is *not* the oldest, both the truther and the liar will say no. Therefore the second question is sufficient to decide if the girl with the necklace is the oldest.

If the robot on the left answers no to the first question, then the robot on the right must be either a truther or liar. The second question is then directed to her, with the same result.

"I just thought of a better solution to your first problem," said Isomorph. "I can learn the identity of all three girls, no matter how they sit, by asking just *two* questions."

What does Isomorph have in mind?

The board game is equivalent to an old game that Henry E. Dudeney, in *Amusements in Mathematics*, problem 392, calls "The Pebble Game." Fifteen pebbles are placed on a table. Players take turns taking 1, 2, or 3 pebbles. After all pebbles are taken, the person who holds an odd number of pebbles wins.

It is easy to see the isomorphism. Not counting the two starting spots, there are 15 spots on the board. Each time a player advances his piece it is the same as removing 1, 2 or 3 spots from those that remain between the two pieces. When the counters meet, all the spots (pebbles) are gone, and the player with his counter inside the circle has "taken" an odd number of spots.

The game can be generalized to any odd number of pebbles and the taking of any number of pebbles from a through b, where a and b are any positive integers, and b is equal to or greater than a.

Why will the game not generalize to an even number of pebbles?

16

Magazines, like books, have even numbers on the left pages, odd numbers on the right. Therefore pages 27 and 28 are opposite sides of the same leaf. The number of missing leaves is five.

Several days later, when I went back to the shop to complain, I couldn't find it. I swear that at the spot where it had been the two brownstone houses on either side were smack against one another. When the check came back to me I saw that it was endorsed "Raymond Dero Palmer."

POSTSCRIPT

The first problem, which seems impossible to solve because of insufficient data, was devised by Søren Hammer Jacobsen of Denmark. With his permission I published it for the first time in this puzzle.

Raymond Dero Palmer is Raymond A. Palmer, SF writer and editor, who died in 1977. I changed his middle name to Dero to honor his famous hoax about the evil deros who live underground. During the forties, when Palmer was editor of *Amazing Stories* magazine, he ran articles on the deros by Richard Shaver, presenting them as fact, not fiction. Thousands of readers took the deros seriously. For some details about this hoax see the entries on Palmer and on Shaver in *The Science Fiction Encyclopedia*, edited by Peter Nicholls, and my *Fads and Fallacies in the Name of Science*, especially chapter 5 on flying saucers.

Palmer played a major role in launching the flying saucer mania. He also founded *Fate* magazine, which continues to publish worthless articles on every aspect of pseudoscience and the paranormal, and advertisements for paranormal junk.

Couth had forgotten to make clear that the impossibility proof assumes that the figure is drawn on a plane. Tanya, who had a high I.Q. for mathematics, realized that the proof fails if the figure is drawn on the surface of a torus (the surface of a doughnut or bagel) in such a way that the hole in the torus is inside one of the compartments surrounded by five line segments. For example, if the figure is chalked inside the *Bagel* as shown below, it is solved easily:

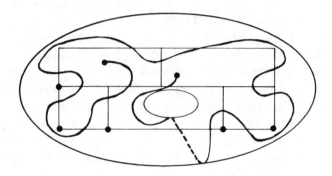

Couth was delighted by his daughter's insight. He then gave her another problem which she solved quickly.

"Suppose," said Couth, "our spaceship were a sphere instead of a torus. Can the figure be drawn on the inside of a sphere in such a way that the original problem can be solved?"

18

Say to a friend: "Will the next word you speak be 'no'? Please answer by saying 'yes' or 'no.'"

POSTSCRIPT

I do not know who first thought of the red-and-green-light version of the computer prediction paradox. It was the basis for a variation I introduced as a betting game in a *Scientific American* column that became chapter 11 of my *New Mathematical Diversions*. For two other prediction paradoxes, each more difficult to resolve than this one, see puzzles 20 and 31 of this book.

Charles Blabbage is an obvious play on Charles Babbage, the British pioneer of computers that can be programmed. Babbage's good friend and disciple, Ada Augusta, the beautiful young countess of Lovelace, was wealthy, witty, intelligent, a good mathematician, and the only legitimate child of the poet Lord Byron. She was the first to say that computers do only what they are told to do. The character of Ada, in Nabokov's novel *Ada*, is partly based on Lady Lovelace.

If you want to know more about this remarkable pair, see *Charles Babbage and His Calculating Engines* by Philip and Emily Morrison; *Ada, Countess of Lovelace* by Doris Langley Moore; and "Byron's Daughter" by B. H. Neumann in the *Mathematical Gazette*, Vol. 57, June 1973, pp. 94–7.

19

Place the 26 black cards of a deck in one pile. Next to it place, say, 13 red cards. Turn your back and ask someone to take as many cards as he likes from the black pile and shuffle them into the red pile. Then take the same number of cards from the former all-red pile and shuffle them into the black pile.

You turn around, massage your temples, and announce that your clairvoyant powers tell you that the number of red cards among the black is exactly the same as the number of black cards among the red.

This must always be the case, and for the same reason given in the solution to the martini problem. If you like, you can let a spectator shuffle the two piles together, then deal 26 cards into one pile and 13 into a second pile. The final result will be the same as before.

Now go back and reread the first part of this feature. What whopping error was made in describing Count Dracula's mixing of the cocktails?

On Thursday morning Philbert was told he would be erased that afternoon. Since Philbert had no way of knowing it would be Thursday, this news came to him as a total surprise. His erasure took place on Thursday. Everything the judge said proved to be accurate.

POSTSCRIPT

This is one of the most notorious of the prediction paradoxes of modern philosophy; for two others, see puzzles 18 and 31. A fuller discussion of the paradox, and a listing of twenty-three papers on it, are found in my book, *The Unexpected Hanging*, chapter 1. Since the book's publication in 1969, more than a dozen new papers have been published. They are listed in an extensive bibliography accompanying a forthcoming paper, "Expecting the Unexpected," by Maya Bar-Hillel and Avishai Margalit.

21

The anagram for *de Camp* is *Decamp*.

If that caught you off guard, here is a marvelous new anagram discovered recently by using an anagram computer program written by one of the hackers in the computer science department at Stanford University. What common eight-letter word, in every English dictionary, is an anagram of *Pictures*?

22

If the square matrix is taken left to right, and top to bottom, the DNA message fills it as shown below:

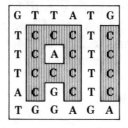

Note that the **C** letters, when shaded, depict the initials **A.I.**

"Curiouser and curiouser!" exclaimed Crock. "What does **A.I.** mean? Artificial Intelligence? If so, it might be a way of saying the virus was made and sent here by robots? But how could they know our English alphabet?"

"I just had a thought," said Witson. "Scanning left to right is no more than a cultural convention. Both letters are mirror symmetrical. If we scan the same matrix right to left we get this." Witson quickly sketched the matrix, lettered its cells, and shaded the **C**s to produce the matrix on the next page.

"So what does **I.A.** stand for?" asked Crock. "Isaac Asimov?"

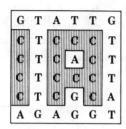

The two men spent several days testing the sequence for other clues, trying matrices of 2×18, 3×12 and 4×9, scanning them in different ways, but no other patterns or number sequences turned up. A few weeks later the entire mystery was unraveled. Can you guess what it was?

Like the previous problem, this can be solved the hard way by drawing internal construction lines and applying algebra. The aha! way is to inscribe another equilateral triangle as shown below:

23

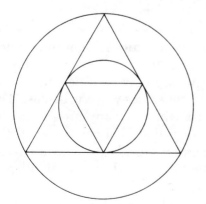

You see at once that the small triangle's area is one-fourth that of the large triangle. Since the small circle and its inscribed triangle are simply the large circle and its triangle reduced in size, the circles must be reduced by the same proportion as the triangles. Therefore the small circle's area is one-fourth that of the large circle.

Suppose the symbol had been an ellipse with maximum-area isosceles triangles inscribed and circumscribed with parallel sides as shown below:

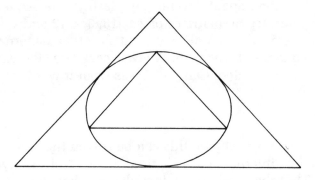

What is the ratio of the areas of the two triangles?

24 Dr. Loveface had read some old SF stories about time travel into the past that got around the familiar difficulty of whether a person would or wouldn't exist if he entered his past and killed his parents when they were babies. The gimmick was to assume that whenever anything from the future enters the past in a way that changes the past, the universe splits into two parallel worlds that are identical except that in one the alteration took place, in the other it didn't.

This gimmick can be applied to tachyonic messages. As soon as such a message enters the universe's past, the big split occurs. A person sending such a message can never get a reply because he continues to exist in the universe in which the message was sent, not in the duplicate universe in which the message was received. This permits one-way communication without contradiction, but not an exchange of tachyonic messages within the same universe.

POSTSCRIPT
The tachyon telephone is closely related to time travel paradoxes. You'll find these paradoxes discussed, together with the telephone, in my *Scientific American* column for May 1974. On the telephone paradox see "The Tachyonic Antitelephone," by G. A. Benford, D. L. Book, and W. A. Newcomb (we will meet Newcomb again in puzzle 32), in *Physical Review D*, vol. 2, July 15, 1970, pp. 263–65. See also the excellent entries on "Tachyons," "Time Travel" and "Time Paradoxes" in *The Science Fiction Encyclopedia*, edited by Peter Nicholls.

An excellent introduction to tachyon theory is Gerald Feinberg's article on "Particles that Go Faster than Light," *Scientific American*, February 1970. Some parapsychologists have suggested that tachyons (if they exist) might be carriers of precognitive ESP (if it exists).

There is no such sequence. This is easily proved by following the same construction procedure used earlier. When you reach this point:

```
1  3  9 . . .
  2  6 . . .
    4  9 . . .
      5 . . .
```

you encounter an unavoidable duplication of 9.

Although it has not yet been proved, it looks as though

Hofstadter's original weird sequence cannot be generalized beyond one row of differences. David J. Bell, an IASFM reader from Campbell, California, pointed out that weird sequences can be generalized to more than one row of differences if we drop the rule that the first numbers of each row must form an increasing sequence.

Now see if you can guess why the *Bagel*'s computer is called VOZ.

27

The bet favors Lucifer again. One ball picks out a slot, and the chances the other ball will find the same slot are 138, not 1 1, 444. Since the correct odds are 37 to 1 that the balls fall in different slots, odds of 100 to 1 give the Devil a sizeable long-run edge.

Even though the bet favored him, the Devil was so infuriated when he lost that he conjured up all his psi energy to lay a curse on the student that made it impossible for him to win any bet in the casino for the rest of the night. Then the Devil, in a huge huff, hied himself back to Hell to get a decent night's sleep.

Suzie, the student's girl friend, was almost as psychic as Satan himself. She realized that the swarthy stranger was a man of limited psi powers, and she sensed at once the nature of his curse. Nevertheless she was delighted by the spell. Why?

The minute hand goes twelve times as fast as the hour hand. As we learned in puzzle 11, the two hands coincide eleven times during every twelve-hour period. Therefore the time between any two consecutive coincidences will be $\frac{12}{11}$ or 65 and $\frac{5}{11}$ minutes, which is the same as 65 minutes, 27 and $\frac{3}{11}$ seconds. The answer to the question is that after 12 o'clock the two hands will again meet precisely at 5 minutes, 27 and $\frac{3}{11}$ seconds past 1 o'clock.

Back to lunar shuttles. Each ship travels a constant speed of 20,000 mph. On one tragic occasion, when two ships were exactly 83,000 miles apart and going opposite ways, the radar system on one craft malfunctioned. This put the ships on a straight-line collision course. How far apart were they fifteen minutes before they crashed?

The paragraph contains every letter of the alphabet except *e*, the last letter of "time."

Now go back and study the original narrative. Somewhere in the text is a block of letters which taken forward spell the last name of a top science fiction author who has written about time travel. There may be spaces between letters, as for example in the word "fat" that is hidden in the second sentence of this paragraph. After you find the last name, look for another sequence of letters in the narrative which taken backward spell the same author's first name. What is the full name?

31

Experts disagree! Some favor the "pragmatic argument" (take only the opaque box). Some favor the "logical argument" (take both boxes). Some say the paradox is not yet resolved. Still others maintain that the paradox proves the impossibility of prediction machines that work with better than 50 percent accuracy.

The problem is known as "Newcomb's paradox" after the American physicist William A. Newcomb who invented it in 1960. Zonick is an anagram of Nozick. It was Robert Nozick, a philosopher at Harvard, who first wrote about the paradox, and who contributed a guest column on it to the Mathematical Games department of *Scientific American* (see the fifth entry in the postscript's list of references). Nozick's column surveys the thousands of letters from readers, including one from Asimov, who sought to resolve the paradox after I discussed it in an earlier column.

POSTSCRIPT

If you care to read some of the growing literature on this bewildering paradox, here is a chronological list of selected references:

Nozick, Robert. "Newcomb's Problem and Two Principles of Choice." *Essays in Honor of Carl G. Hempel*, edited by Nicholas Rescher, 1970.

Howard, Nigel. *Paradoxes of Rationality: Theory of Metagames and Political Behavior*, 1971, pp. 168–84.

Bar-Hillel, Maya and Avishai Margalit. "Newcomb's Paradox Revisited." *British Journal for the Philosophy of Science*, vol. 23, November 1972.

Gardner, Martin. "Free Will Revisited." Mathematical Games Department, *Scientific American*, July 1973, pp. 104–109.

Nozick, Robert. "Reflections on Newcomb's Problem." Mathematical Games Department, *Scientific American*, March 1974, pp. 102–107.

Schlesinger, G. "The Unpredictability of Free Choices." *British Journal for the Philosophy of Science*, vol. 25, September 1974.

Levi, Isaac. "Newcomb's Many Problems." *Theory and Decision*, vol. 6, May 1975.

Brams, Steven J. "Newcomb's Problem and Prisoners' Dilemma." *Journal of Conflict Resolution*, vol. 19, December 1975.

Brams, Steven J. "A Paradox of Prediction." *Paradoxes in Politics* (ch. 8), 1976.

Locke, Don. "How to Make a Newcomb Choice." *Analysis*, vol. 38, January 1978.

Lewis, David. "Prisoners' Dilemma Is a Newcomb Problem," *Philosophy and Public Affairs*, vol. 8, Spring 1979.

32

The ostrich proposed that the clerk move the occupants of each room n to a room of number $2n$. Those in 1 went to 2, those in 2 to 4, those in 3 to 6, and so on. This vacated every room with an odd number. Since there is an infinity of odd numbers, the SF fans (all of whom *were* rather odd) were easily accommodated.

POSTSCRIPT

If you wish to learn more about the paradoxes associated with aleph-null (the number of the set of all positive integers), and about Cantor's higher alephs, a good place to start is the second chapter of *Mathematics and the Imagination*, by Edward Kasner and James Newman. See also the chapter on "Aleph-null and Aleph-one" and the bibliography in my *Mathematical Carnival*.

34

The abbreviation of Pennsylvania is PA. Assuming that the alphabet is cyclical, so that A and Z are joined, shifting each letter in PA *back* one step also arrives at OZ!

Dot and Tot of Merryland was the first fantasy novel Baum wrote after the huge success of his *Wizard of Oz*. Can you think of a plausible reason why he named the hero and heroine of this novel Dot and Tot?

35

"The number is 54," said Zero. "The divisors that catch all nine digits (excluding 0) are 1, 2, 3, 54, 6, 27, 18, 9. The smallest number whose divisors catch all ten digits is 108, or twice 54. We now conclude our performance with our *pièce de résistance*." He raised one hand, shouted the mysterious word "Erdös!" and instantly his fingers closed around a large parchment scroll that appeared from nowhere. He unrolled it to display the following strange diagram:

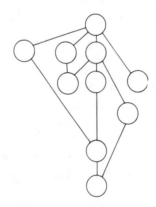

"In each circle," Zero explained, "you are asked to place one of the ten digits. All ten must be used. 'Smaller' digits are below 'larger' digits, but what is meant by smaller and larger is left undefined. The lines stand for a binary relation that each digit has with any lower digit to which it is joined."

"I'll never remember that diagram," I recall mumbling in my sleep.

"In that case," said Zero, "I'll leave it with you."

He leaped into the air, turned a back flip, and landed on his feet with a thunderclap. All ten digits vanished in a cloud of emerald smoke. I woke up, snapped on the light, and found the scroll on the rug. When I unrolled it, the digits were properly inscribed on it with green ink. Can you figure out the meaning of the diagram?

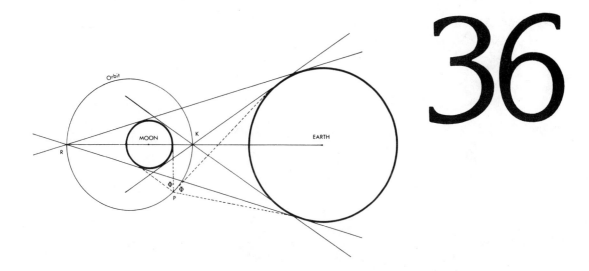

Draw the two external tangents as shown in the picture. They intersect at R. The earth and moon will appear the same size at both K and R. The large circle on which points R and K lie is the orbit from which the earth and moon will always seem the same size. In other words, from any point P on this circle, the pair of angles labeled with the Greek letter *phi* will be equal. It is called the "circle of similitude." You can find out more about its properties in textbooks on plane geometry.

Have you ever wondered about the curious fact that viewed from the earth the sizes of the sun and moon are almost identical? That is why the disk of the moon almost exactly covers the disk of the sun during a total solar eclipse. Can you think of any good reason why this should be the case?

We learned in the first answer that the proportions of male, female, and bisexual children remain permanently 1:1:1. Because of the decree every mother has exactly one bisexual child. To preserve the 1:1:1 ratios, the average number of males to a mother must be 1, and the average number of females must also be 1. This makes an average of 3 children per mother.

POSTSCRIPT

The Byronia problem is my generalization of "A Family Problem" that I found in *Puzzle-Math*, a collection of delightful problems by George Gamow and Marvin Stern.

In case you wonder how to find the limit of the infinite sequence $1 + \frac{2}{3} + \frac{4}{9} + \frac{8}{27} + \ldots$, here is a simple way to do it.

Let x stand for the sequence with the first term omitted. In other words,

$$x = \frac{2}{3} + \frac{4}{9} + \frac{8}{27} + \ldots.$$

Each term is $\frac{2}{3}$ the previous one. Multiplying both sides by $\frac{2}{3}$ produces:

$$\frac{2x}{3} = \frac{2}{3}\left(\frac{2}{3} + \frac{4}{9} + \frac{8}{27} + \ldots\right)$$

$$\frac{2x}{3} = \frac{4}{9} + \frac{8}{27} + \frac{16}{44} + \ldots.$$

The sequence on the right is $\frac{2}{3} + x$. We can now write:

$$x = \frac{2x}{3} + \frac{2}{3}$$

which, when solved, gives x a value of 2. Since x is the original sequence minus 1, we add 1 to obtain 3 for the limit sum of the sequence.

6

If the process starts with 2 microbes, then after each hour the number will be double the corresponding number in the sequence that starts with 1 microbe. After 48 hours, twice the number of microbes is not enough to make the container $\frac{1}{7}$ full, so the correct answer is 49 hours. At that time the container becomes $\frac{2}{7}$ full.

POSTSCRIPT

The second and third variants of this problem are new. The second variant is a good problem to try on friends because most people divide 49 by 7 to get an answer that is way off. Readers familiar with Wells's story will recognize Montgomery as the name of Dr. Moreau's assistant.

7

On any graph, a "loop" (a line that joins a point to itself) adds *two* more lines to the point. Therefore, it has no effect on the point's evenness or oddness.

POSTSCRIPT

The story line for this problem is based on a joke I first heard from the Canadian mathematician Leo Moser, noted for his quickness of mind, his originality, his expertness at chess, and his large collection of mathematical jokes, limericks, and cartoons. However, because *IASFM* is a family magazine, the subject of the original joke was considered inappropriate.

The editor prudently changed the binary relation to one of shaking hands.

Lieutenant Flarp's statement, which begins "How I want a drink," is a mnemonic for pi devised by the famous British astronomer Sir James Jeans. The number of letters in each word corresponds to the first fifteen digits of pi.

8

POSTSCRIPT

Many elegant problems in mathematics can be solved quickly if you make the assumption that the problem does indeed have a solution. A classic example, in three dimensions, closely related to this one, concerns a cylindrical hole that has been drilled straight through the center of a solid wooden sphere. The hole is six inches long. What is the volume remaining in the sphere?

The problem can be solved the hard way, but let's take the shortcut of assuming there is a solution. If so, the volume must be a constant regardless of the hole's size. Reduce the radius of the hole to zero. What "remains" is a solid sphere with a diameter of six inches and a volume of thirty-six times pi. If the problem has a unique answer, this must be it!

Dozens of clever mnemonic sentences for remembering decimal expansions of pi and other irrational numbers have been published. Here is a short nonalcoholic sentence for pi to ·seven decimal places: *May I have a large container of coffee?*

The two other anagrams are TRACER and RE-CART.

12

13

1. The only anagram of GREEN is GENRE.

2. Lewis Carroll (who invented this word game) changed PINK to BLUE in nine steps as follows: PINK, PINT, PANT, PART, PORT, POUT, GOUT, GLUT, GLUE, BLUE.

By using one or two less familiar words it can be done in seven steps. Here is an example, using words that all appear in *Webster's New Collegiate Dictionary*: PINK, PINT, PENT, PEAT, BEAT, BLAT, BLAE, BLUE.

POSTSCRIPT

I confess that I must be blamed for the first of these problems. If you want some more examples of Lewis Carroll's word ladders (he called them "doublets"), see the chapter on Carroll in my *New Mathematical Diversions from Scientific American*. Donald Knuth, a computer scientist at Stanford University, has devised a computer program that finds minimum-length word ladders in microseconds.

14

Say to any robot: "If I were to ask each of you whether you are male or female, and your two companions gave the same answer, would your answer agree with theirs?"

The truther would have to say no, the liar would have to say yes, and the sometimer would be unable to reply because she knows that her companions (one a truther, the other a liar) could not give the same answer. By directing this curious question toward any two robots, their identities are established and you will know the identity of the third.

"I must admit," said Isomorph, "that the sometimer could answer yes or no, and either answer would be, in a vague sense, a lie. But I assume that the question would at least cause the sometimer to think a long time before answering, if at all. Therefore I maintain this is a legitimate two-question solution to the first problem."

POSTSCRIPT

Oz fans will recognize that Professor Tinker's full name honors both Lyman Frank Baum, who wrote the first series of Oz books, and Mr. Tinker, of the firm of Smith and Tinker. It was Tinker who invented and constructed Tiktok, one of the earliest mechanical robots in American fiction. Baum introduced Tiktok in his third Oz book, *Ozma of Oz*, in which Dorothy is shipwrecked and washed ashore on a beach of Ev, a magic land adjoining Oz. Of course Asik Isomorph honors Isaac Asimov.

Puzzle literature is filled with problems involving truthers, liars, and sometimers. For problems related to the ones given here, see my *Sixth Book of Mathematical Games from Scientific American*, chapter 20; *Journal of Recreational Mathematics*, vol. 12, no. 4, 1979–80, p. 311; and C. R. Wylie, Jr., *101 Puzzles in Thought and Logic*, problem 47.

Aaron J. Friedland wrote to suggest a single question that can identify the truther, the liar, the sometimer, and the oldest robot, as well as all other desired information known to all three. The question, directed to any robot, is: "If I were to ask 'Which of you is the truther, which the liar, which the sometimer, and which the oldest?' and your answers were as truthful as your answers to the present question, what is a possible set of answers you could give?" Friedland used this form of question to show how to obtain a truthful answer from a sometimer in his book *Puzzles in Math and Logic*, puzzle 99.

15

If the number of pebbles at the start of a game is even, the game must end in a draw no matter how it is played. No even number can be the sum of an even and odd number, therefore a game beginning with an even number of pebbles will end with both players holding an odd number, or both holding an even number.

It is, of course, possible to draw the pattern so that one player starts inside the circle, the other outside, and between the starting points there is an even number of spots. This is the same as playing the pebble game with an even number of pebbles, and giving one player a single pebble at the start of the game. It does not change the basic character of the game.

POSTSCRIPT

The name Sidney Bagson plays on the name of an old friend from New York City, Sidney Sackson. A board game collector and game inventor extraordinaire, he is the author of *A Gamut of Games* and other books on games. The game described here was devised by Frank Tapson of Exeter, England. A slightly different version of the game appears in his book *Take Two! 32 Board Games for Two Players.*

The generalized pebble game is discussed in Geoffrey Mott-Smith's *Mathematical Puzzles*, problem 177; and Roland Sprague's *Recreation in Mathematics*, problem 24.

17

The problem is not solvable on a sphere's surface. Assume there is a solution, and that the sphere is a rubber sheet. Puncture it at any point not on a line of the figure or on the line that solves the puzzle. The punctured sphere can now be stretched to make a plane surface. Since this stretching does not alter the topological properties of the figure, it would produce a solution on the plane. As we have seen, however, there is no solution on the plane. Consequently there cannot be a solution on the sphere.

POSTSCRIPT

The original problem, of unknown origin, is one of the oldest and most popular of all topology puzzles. Like Sam Loyd before me, I am constantly getting letters from readers who are unable to crack it. The proof that it has no solution is one of the simplest, most elegant examples of how a "parity check" (a check based on odd and even numbers) can quickly answer a question that otherwise would be difficult.

To my knowledge, I was the first to note the fact, albeit trivial, that the problem can be solved on a torus. I gave this solution in a 1957 *Scientific American* column that became chapter 12 of my *Scientific American Book of Mathematical Puzzles and Diversions*. Ronald Couth is a play on Donald Knuth, the distinguished computer scientist mentioned in my postscript to puzzle 13.

19

The description said that Mrs. Dracula watched her husband in a mirror. As every reader knows, or should know, vampires don't *have* mirror reflections.

POSTSCRIPT
Hundreds of mathematical card tricks exploit essentially the same principle as the one just described. Here is a good example to try on friends.

Before showing the trick, cut a deck of 52 cards exactly in half. Turn over one half, then shuffle the 26 face-up cards into the 26 face-down ones. When you start the trick, show that the deck is a mixture of face-up and face-down cards, but don't say how many are reversed. Let someone shuffle, then hand you the deck under a table. A moment later you bring out the cards, a half-pack in one hand, the other half in the other, and announce that each half contains exactly the same number of face-up cards! This proves to be correct.

Secret: Under the table, count off 26 cards. Turn over either of the half-packs before you put the two packets of cards on the table. Do you see why it works? Before reversing one half, the number of face-up cards in either half must equal the number of face-down cards in the other. Reversing either half turns its face-up cards face down and vice versa. This makes the number of face-up (or face-down) cards the same in each half-deck.

21

Piecrust.

POSTSCRIPT

To find out more about the Oulipo look up my February 1977 Scientific American column on this slightly mad group and check the references cited at the back of the magazine.

Edward I. Schmahl wrote that he, too, had been unable to find an English anagram for Asimov, but he did find an ungrammatical two-word French anagram: *vos ami* (your friend).

John H. Shurtleff, a Chicago patent lawyer, applied to piecrust a popular anagrammatic game in which you keep adding letters and rearranging to form new words. Thus *piecrust + R = scripture. Scripture + I = cruise trip.*

A variation of the game is to combine words with words. Thus *piecrust + eon = persecution. Piecrust + eon + earth = terpsichorean ute.*

The technique of adding a single letter to a word, then rearranging the letters to make a new word, is known as "transaddition" (or transdeletion if you chop out letters). Dmitri Borgmann, in his classic *Language on Vacation*, gives three remarkable transaddition chains: *A* to *concentrations* (*a, at, tea, sate, . . .*); *O* to *precariousness* (*O, no, eon, nose, . . .*); and *I* to *semipardonable* (*I, is, sir, rise, . . .*).

The virus was artificial all right, but not from outer space. A team of biologists at the Artificial Intelligence Laboratory of nearby M.I.T. was engaged in a secret military project to determine if it were possible to transmit cryptographic information by artificial viruses. The project was headed by Isaac Asimov III, a descendant of the celebrated SF and S writer. He had used his own initials in the DNA sequence, aware that they could be reversed to stand for Artificial Intelligence. An assistant had dropped a crock containing the virus. The crock cracked, specimens of the virus escaped the laboratory, and were airborne to the Harvard Yard.

22

POSTSCRIPT
The two Tokyo scientists mentioned in this puzzle are real. They are Hiromitsu Yokoo and Tairo Oshima. Their conjecture that viruses might carry coded messages from an extraterrestrial civilization was published in the April 1979 issue of Icarus, an international journal of astronomy devoted to studies about the solar system.

According to Walter Sullivan's account of their work (in the New York Times, May 7, 1979), Yokoo and Oshima were led to their curious theory by the discovery that the genetic sequence in a bacteria-infecting virus called PhiX-174 seems more contrived than natural.

As explained in my problem, genetic information is coded along the DNA molecule in the form of three-letter "words" that use a four-letter "alphabet." The Japanese scientists suggested that extraterrestrial intelligences might find it easier to send coded messages by dropping viruses on a planet, where they would multiply rapidly, than by using radio signals. No messages were found in PhiX-174, but the authors recommend that similar efforts be made to look for coded messages in other viruses.

23

Tanya's trick of inverting the inside triangle works here, too, although it is not obvious that the upper corners of the inverted triangle will touch the ellipse at spots where the curve is tangent to the large triangle's two sides.

However, if you are familiar with affine geometry you will recall that affine "stretching" or "shrinking" of a figure preserves all area ratios. Shrinking the ellipse horizontally until its two foci merge will change it to a circle with inscribed and circumscribed equilateral triangles. This is the original Titan symbol. Since area ratios are preserved by stretching the circle to an ellipse, the ratio of the areas of the two isosceles triangles is 1 to 4 as before.

POSTSCRIPT

I do not know the origin of the first problem. The second came to me from Daniel R. Royalty, of Ames, Iowa. The third variant is my own. Larc Snaag, by the way, is an anagram of Carl Sagan, and "Scitheration" derives from George Scithers, the able editor of *IASFM*.

At this time, Titan seems to be the most likely body in

the solar system to have life. Unfortunately, the *Voyager 1* spacecraft that approached Saturn in 1980 failed to learn much about Titan, but let us hope that later probes will.

26

Remember HAL, the computer in the movie *2001*? If you shift each letter of HAL one step forward in the alphabet you get IBM, whose logo is clearly visible on the computer in the film. Shift each letter in VOZ thirteen steps (regard the alphabet as cyclical) and you also get IBM.

Since thirteen is half of twenty-six we can describe the transformation in a more dramatic way. Write the twenty-six letters of the alphabet in a circle, then for each letter of VOZ substitute the letter diametrically opposite.

POSTSCRIPT

This puzzle has been considerably revised since it was first published in *IASFM*. I was then unaware that the doubly-weird sequence duplicated the number 284. Karl Fox, of Columbus, Ohio, wrote me with this new insight. No duplication is possible in the original sequence because each number in the first row of differences is chosen by taking the smallest integer not yet used. It may be a difficult task to prove there are no higher-order sequences, or to find one or more such sequences.

27

Suzie instructed her friend to play the colors with single chips, each worth 1 ozmuf. Every time he placed a bet on red or black, Suzie put 100 ozmufs on the other color!

POSTSCRIPT

The first problem is based on one of the most counterintuitive of many nontransitive paradoxes. If a binary relation that links A to B, and B to C, necessarily links A to C, it is called transitive. For example, if A is taller than B, and B is taller than C, then A must be taller than C, therefore the relation "taller than" is transitive.

Intuition tells us that if triplet A beats triplet B and triplet B beats triplet C, then triplet A should beat C. In this case, however, the relation is nontransitive so it doesn't work that way. For more details on this amazing betting game, see my column on nontransitive paradoxes in *Scientific American*, October 1974.

David VomLehn of Hanover, New Hampshire, sent the following sequel to the story of Lucifer at Vegas: In the early hours of the morning, the Harvard student sensed that something mighty strange was going on. To test his suspicions he made a side bet with a gambler who was playing the same roulette wheel. As the ball settled in one of the slots, the entire universe suddenly stopped moving. Why?

VomLehn explains that the student had bet that he would lose his next roulette bet. Thus, whichever bet he lost, he would be certain to win the other. The Devil's curse covered all bets for the night. Faced with an unresolvable logical contradiction, the universe automatically halted!

It is unnecessary to know the distance between the two ships when the radar system broke down. That figure was given only to distract you from the ridiculously simple solution.

The two ships have an approach speed of 2 times 20,000 mph or 40,000 mph. In your mind, run the scene backward in time from the crash. An hour before colliding they must have been 40,000 miles apart. Fifteen minutes before colliding they would have been one-fourth that distance, or 10,000 miles, apart.

POSTSCRIPT

Here are the exact times that the hour and minute hands meet during every twelve-hour period:

12: 00: 00
  1: 05: 27 and $3/11$
  2: 10: 54 and $6/11$
  3: 16: 21 and $9/11$
  4: 21: 49 and $1/11$
  5: 27: 16 and $4/11$
  6: 32: 43 and $7/11$
  7: 38: 10 and $10/11$
  8: 43: 38 and $2/11$
  9: 49: 05 and $5/11$
10: 54: 32 and $8/11$

Many clocks also have sweep second hands. How many times during a twelve-hour period do all *three* hands coincide? The surprising answer is: only at 12 o'clock! You will find two proofs of this on pages 60–1 of my *Sixth Book of Mathematical Games from Scientific American*.

29

The author is Isaac Asimov. Both names are in the sequence of capitalized letters in the sentence that starts: "BianCA, AS I MOVe these levers. . . ."

POSTSCRIPT

Palindromes provide an enormous area of exploration for both wordplay enthusiasts and number theorists. For a discussion of palindromes of both types see chapter 19 and the bibliography of my book, *Mathematical Circus*. For word palindromes there is no better guide than Howard Bergerson's paperback, *Palindromes and Anagrams*.

34

In *Dot and Tot of Merryland*, Dot could stand for Dorothy and Tot could stand for Toto.

POSTSCRIPT

For more scrambled verse see "Pied Poetry," Martin Gardner and J. A. Lindon, *Word Ways*, vol. 6, May 1973, pp. 98–100.

No one really knows why Baum chose the name Oz. The legend that Baum looked up and saw O-Z on a filing cabinet first appeared in the *New York Mirror*, January 27, 1904. Reasons for doubting the truth of this are given in *The Baum Bugle*, Spring 1969, a quarterly journal which contains a wide variety of material pertaining to the Oz series. The forward shift of OZ to NY was discovered by Mary Scott of the International Wizard of Oz Club that publishes this periodical. I found the backward shift of OZ to PA.

Northwest of Oz is the huge relatively unexplored land of Ev, the home of the evil Nome King. In choosing Ev, Baum may have had in mind the first two letters of "evil." Michael Hearn, author of *The Annotated Wizard of Oz*, was the first to notice the link between Dot and Tot, and Dorothy and Toto.

The solution is shown below. The digits are drawn with bars as on a calculator readout. All seven bars are used for 8 at the top; six bars for 0, 6, 9; five for 5, 3, 2; four for 4; three for 7; and two for 1. Each digit is obtained by removing one or more bars from any higher digit to which it is joined by a line.

35

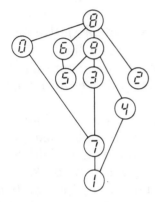

POSTSCRIPT

For a complete listing of all squares formed with digits 1 through 9, and digits 0 through 9, see "Squares with 9 and 10 Distinct Digits," by T. Charles Jordan, in *Journal of Recreational Mathematics*, vol. 1, January 1968, pp. 62–3.

My information about the products of twins and triplets is from a 1972 letter from Y. K. Bhat, of New Delhi. He found seven solutions for nine digits, nine solutions for ten digits:

$$28 \times 157 = 4396$$
$$27 \times 198 = 5346$$
$$18 \times 297 = 5346$$
$$42 \times 138 = 5796$$
$$12 \times 483 = 5796$$
$$39 \times 186 = 7254$$
$$48 \times 159 = 7632$$

$$39 \times 402 = 15678$$
$$27 \times 594 = 16038$$
$$54 \times 297 = 16038$$
$$36 \times 495 = 17820$$
$$45 \times 396 = 17820$$
$$52 \times 367 = 19084$$
$$78 \times 345 = 26910$$
$$46 \times 715 = 32890$$
$$63 \times 927 = 58401$$

Clement Wood, on page 48 of *A Book of Mathematical Oddities* (Little Blue Book No. 1210), published in Girard, Kansas, by Haldeman-Julius, undated), points out that $27 \times 594 = 16038$ is unique in that one factor is a multiple of the other ($594 = 27 \times 22$).

Incidentally, the Little Blue Books was an innovative and provocative series. The books were published by an American socialist and freethinker, Emmanuel Haldeman-Julius, and were forerunners of today's paperbacks. The books measured 3½ by 5 inches, and sold by the hundreds of millions during the twenties and thirties for five cents each. There were thousands of titles.

The seven solutions for the nine positive digits are given on page 65 of Albert Beiler's *Recreations in the Theory of Numbers*. He adds the following equalities:

$$4 \times 1738 = 6952$$
$$4 \times 1963 = 7852$$
$$3 \times 51249876 = 153749628$$
$$6 \times 32547891 = 195287346$$
$$9 \times 16583742 = 149253678$$

Is it possible to group the nine positive digits into three triplets that have a product containing the same nine digits? Yes, though I do not know the total number of solutions. Hilario Fernandez Long, of Argentina, worked by hand on this problem in 1972. The highest product he found is 435918672, which equals $567 \times 843 \times 912$. The lowest product he found is $127386945 = 163 \times 827 \times 945$. His solution with the highest triplet is 964 × 738 × 251 = 178569432. His solution with the lowest triplet is the one given above as having the lowest product. Note the curious fact that the triplet 945 reappears at the end of the product.

An old curiosity involving the nine positive digits—it goes back to medieval times—is:

```
  987654321
 −123456789
  864197532
```

All nine digits appear in the difference! Astonishingly, the same thing happens if all ten digits are used.

```
  9876543210
 −0123456789
  9753086421
```

The problem answered by 54 is my own, though I based it on a question sent to me in 1980 by Bruce Reznick, of the University of Illinois, at Urbana. Reznick asked for the smallest digit n which has positive divisors (including n and 1) *ending* in every digit, and showed that the answer is 270. If we change this to divisors that *begin* with every digit except 0, the answer is 216. Note that all four answers—54, 108, 216 and 270—are multiples of 54.

The problem with the diagram was sent to me in 1980 by its inventor, Raphael M. Robinson, a distinguished mathematician at the University of California, Berkeley.

Why did not the jolly green digits form the lowest and highest prime numbers containing all nine positive digits (with or without zero)? The answer is simple. There *are* no such numbers. The nine digits add to 45, and $4 + 5 = 9$. Thus 9 is the digital root of any number formed by permuting the nine digits, and of course a zero anywhere in the number will not alter its digital root. Because any number with a digital root of 9 is evenly divisible by 9, the number cannot be prime.

36

There is no known reason why the sun and moon should appear so nearly the same size from the earth. Astronomers regard it as sheer coincidence. There are so many millions of ways that remarkable coincidences like this can turn up in astronomy that a certain number should be expected. The sun-moon disk equality is simply one of them.

POSTSCRIPT

There is another remarkable coincidence about the sun and moon. The sun's surface rotates at different rates—slower at the poles, faster at the equator. An observer on the moon would see the sun's equator make one turn in about twenty-eight days. To an observer on the sun, the moon also rotates once in about twenty-eight days!

The second problem, finding the orbit that is a circle of similitude, is based on a problem solved in *Mathematics Magazine*, vol. 41, May 1968, p. 133. Charles E. Maley had posed: "Construct an orbit of a spaceship such that the moon and earth will always appear equally large to the astronauts."

Bibliography

Baum, L. Frank. *Dot and Tot of Merryland*. Chicago: George M. Hill, 1901.

———. *Ozma of Oz*. New York: Ballantine Books, 1979.

———. *Tik-Tok of Oz*. New York: Ballantine Books, 1980.

Beiler, Albert. *Recreations in the Theory of Numbers*. New York: Dover, 1964.

Bergerson, Howard. *Palindromes and Anagrams*. New York: Dover, 1973.

Borgmann, Dmitri. *Language on Vacation*. New York: Charles Scribner's Sons, 1965.

Brams, Steven J. *Paradoxes and Politics*. New York: Free Press, 1976.

Carroll, Lewis. *Sylvie and Bruno Concluded. The Complete Works of Lewis Carroll*. New York: Modern Library, 1939.

Dudeney, Henry Ernest. *Amusements in Mathematics*. New York: Dover, 1958.

Fézandié, Clement. *Through the Earth*. New York: Century Co., 1898.

Friedland, Aaron J. *Paradoxes in Mathematics and Logic*. New York: Dover, 1979.

Gamow, George and Marvin Stern. *Puzzle-Math*. New York: Viking Press, 1958.

Gardner, Martin. *Fads and Fallacies in the Name of Science*. New York: Dover, 1957.

———. *The Scientific American Book of Mathematical Puzzles and Diversions*. New York: Simon & Schuster, 1959.

———. *New Mathematical Diversions from Scientific American*. New York: Simon & Schuster, 1966.

———. *The Unexpected Hanging and Other Mathematical Diversions*. New York: Simon & Schuster, 1969.

———. *Sixth Book of Mathematical Games from Scientific American*. San Francisco: W. H. Freeman, 1971.

———. *Mathematical Carnival*. New York: Alfred A. Knopf, 1975.

———. *Mathematical Circus*. New York: Alfred A. Knopf, 1979.

BIBLIOGRAPHY

Hearn, Michael. *The Annotated Wizard of Oz*. New York: Clarkson Potter, 1973.

Hofstadter, Douglas. *Gödel, Escher, Bach: An Eternal Golden Braid*. New York: Basic Books, 1979.

Howard, Nigel. *Paradoxes and Rationality: Theory of Metagames and Political Behavior*. Cambridge, Mass.: MIT Press, 1971.

Kasner, Edward and James Newman. *Mathematics and the Imagination*. New York: Simon & Schuster, 1940.

Kellermann, Bernhard. *The Tunnel*. New York: Macaulay, 1915.

Moore, Doris Langley. *Ada, Countess of Lovelace*. New York: Harper & Row, 1977.

Mordell, Louis Joel. *Diophantine Equations*. New York: Academic Press, 1969.

Morrison, Emily and Philip. *Charles Babbage and His Calculating Engines*. New York: Dover, 1961.

Mott-Smith, Geoffrey. *Mathematical Puzzles*. New York: Dover, 1954.

Nicholls, Peter, ed. *The Science Fiction Encyclopedia*. New York: Doubleday, 1977.

Pendray, Gawain Edwards. *The Earth-Tube*. New York: Appleton, 1929.

Rescher, Nicholas, ed. *Essays in Honor of Carl G. Hempel*. New York: Humanities Press, 1970.

Sackson, Sidney. *A Gamut of Games*. New York: Random House, 1969.

Sprague, Roland. *Recreation in Mathematics*. London: Blackie and Son, 1963.

Tapson, Frank. *Take Two! 32 Board Games for Two Players*. New York: Pantheon, 1979.

Wylie, C. R., Jr. *101 Puzzles in Thought and Logic*. New York: Dover, 1957.

POSTSCRIPT

PUZZLE 5

Several readers generalized the defective doyles problem to *n* doyles. To obtain a set of weights for each *n*, in which the largest weight is minimized, a sequence called the Conway-Guy sequence is used. It is number M1075 in *The Encyclopedia of Integer Sequences*, by N. J. A. Sloane and Simon Plouffe (Academic, 1995). There you will find the curious recurrence that defines the sequence, and a list of references.

The following table gives the sequence *k* as it relates to n (the number of doyles) from 1 through 10.

n	*k*
0	0
1	1
2	2
3	4
4	7
5	13
6	24
7	44
8	84
9	161
10	309

Physicist Richard Hess sent me the results of his computer program that carried the Conway-Guy sequence through $n = 106$.

The simplest way to explain how to obtain the desired number of weights from the Conway-Guy sequence is to give an example. Consider the case of $n = 6$ which I used in my version of the problem. From the sixth number of the sequence, 24, we subtract each of its preceding numbers:

$$24 - 13 = 11$$
$$24 - 7 = 17$$
$$24 - 4 = 20$$
$$24 - 2 = 22$$
$$24 - 1 = 23$$
$$24 - 0 = 24$$

Amazingly, the remainders give the desired number of weights. For another example, let the number of doyles be 10. The subtrac-

POSTSCRIPT

PUZZLE 5 tions from 309, the tenth number of the sequence, give the weights: 148, 225, 265, 285, 296, 302, 305, 307, 308, 309.

Thus if the scale shows an overweight of, say, 567 milligrams, the only way to reach that sum by adding numbers in the set of ten weights is by adding 265 and 302, so you know that cans 3 and 6 hold the overweight doyles.

PUZZLE 6 Richard Hess wrote to tell me that Dr. Moreau's container must have been unbelievably huge (a volume of many cubic miles) to contain all the microbes after fifty hours of their multiplying, or the microbes would have to be smaller than a hydrogen atom.

PUZZLE 8 When I wrote this puzzle tale it was not known whether Neptune, like Jupiter, Saturn, and Uranus had rings. In 1989 Voyager discovered at least 3 rings around Neptune.

PUZZLE 13 For a more complete discussion of Lewis Carroll's doublets see my book on Carroll's mathematical and linguistic recreations, *The Universe in a Handkerchief* (Copernicus, 1996).

PUZZLE 21 My *Scientific American* column on the Oulipo is reprinted in my *Penrose Tiles to Trapdoor Ciphers* (Mathematical Association of America, 1997), followed by a second chapter on the Oulipo not previously published.

PUZZLE 23 Tanya's Aha! solution to the problem, by turning the smaller triangle upside down, reminds me of a similar problem.

How quickly can you prove that the area of the inside square in the figure below is exactly half the area of the larger square? When I

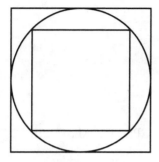

gave this problem to Marilyn vos Savant in 1999, for her *Parade* column, and before she published the answer, I had a long distance phone call from a man who had worked on the problem the hard way. He assured me that the inside square was almost, but not quite, half the area of the outside square. When I told him to rotate the inside square like so:

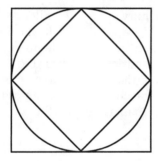

He felt like kicking himself.

H. G. Wells's story about Brownlow's newspaper can be found in **PUZZLE 30** *The Country of the Blind and Other Science-Fiction Stories,* by Wells, a paperback I edited for Dover in 1997. My introduction to the tale gives a more complete list of Wells's hits and misses.

SPECTRUM SERIES

The Spectrum Series of the Mathematical Association of America was so named to reflect its purpose: to publish a broad range of books including biographies, accessible expositions of old or new mathematical ideas, reprints and revisions of excellent out-of-print books, popular works, and other monographs of high interest that will appeal to a broad range of readers, including students and teachers of mathematics, mathematical amateurs, and researchers.

Numerology or What Pythagoras Wrought, by Underwood Dudley

Out of the Mouths of Mathematicians, by Rosemary Schmalz

Penrose Tiles to Trapdoor Ciphers ... and the Return of Dr. Matrix, by Martin Gardner

Polyominoes, by George Martin

Power Play, by Edward J. Barbeau

The Random Walks of George Pólya, by Gerald L. Alexanderson

The Search for E. T. Bell, also known as John Taine, by Constance Reid

Shaping Space, edited by Marjorie Senechal and George Fleck

Student Research Projects in Calculus, by Marcus Cohen, Arthur Knoebel, Edward D. Gaughan, Douglas S. Kurtz, and David Pengelley

Symmetry, by Hans Walser. Translated from the German by Peter Hilton, with the assistance of Jean Pedersen.

The Trisectors, by Underwood Dudley

Twenty Years Before the Blackboard, by Michael Stueben with Diane Sandford

The Words of Mathematics, by Steven Schwartzman

WITHDRAWN

Order MAA publications from:

MAA Service Center

P.O. Box 91112

Washington, DC 20090-1112

800-331-1622   FAX 301-206-9789